The Open University

S342

Science: a third level course

Physical Chemistry

PRINCIPLES OF CHEMICAL CHANGE

BLOCK 5
HETEROGENEOUS CATALYSIS

THE S342 COURSE TEAM

CHAIR AND GENERAL EDITOR

Kiki Warr

AUTHORS

Keith Bolton (Block 8; Topic Study 3)

Angela Chapman (Block 4)

Eleanor Crabb (Block 5; Topic Study 2)

Charlie Harding (Block 6; Topic Study 2)

Clive McKee (Block 6)

Michael Mortimer (Blocks 2, 3 and 5)

Kiki Warr (Blocks 1, 4, 7 and 8; Topic Study 1)

Ruth Williams (Block 3)

Other authors whose previous S342 contribution has been of considerable value in the preparation of this Course

Lesley Smart (Block 6)

Peter Taylor (Blocks 3 and 4)

Dr J. M. West (University of Sheffield, Topic Study 3)

COURSE MANAGER

Mike Bullivant

EDITORS

Ian Nuttall

Dick Sharp

BBC

David Jackson

Ian Thomas

GRAPHIC DESIGN

Debbie Crouch (Designer)

Howard Taylor (Graphic Artist)

COURSE READER

Dr Clive McKee

COURSE ASSESSOR

Professor P. G. Ashmore (original course)

Dr David Whan (revised course)

SECRETARIAL SUPPORT

Debbie Gingell (Course Secretary)

Jenny Burrage

Margaret Careford

Shirley Foster

The Open University, Walton Hall, Milton Keynes, MK7 6AA

Copyright © 1996 The Open University. First published 1996. Reprinted 2002

All rights reserved. No part of this publication may be reproduced, stored in a retrieval system or transmitted in any form or by any means, without written permission from the publisher or a licence from the Copyright Licensing Agency Limited. Details of such licences (for reprographic reproduction) may be obtained from the Copyright Licensing Agency Ltd of 90 Tottenham Court Road, London, W1P 9HE.

Edited, designed and typeset by The Open University.

Printed in the United Kingdom by Henry Ling Ltd, The Dorset Press, Dorchester DT1 1HD

ISBN 0 7492 51867

This text forms part of an Open University Third Level Course. If you would like a copy of Studying with The Open University, please write to the Central Enquiry Service, PO Box 200, The Open University, Walton Hall, Milton Keynes, MK7 6YZ. If you have not enrolled on the Course and would like to buy this or other Open University material, please write to Open University Educational Enterprises Ltd, 12 Cofferidge Close, Stony Stratford, Milton Keynes, MK11 1BY, United Kingdom.

s342block5i1.2

CONTENTS

1	INTRODUCTION	5
2	A DEFINITION	7
3	A SIMPLIFIED POTENTIAL ENERGY PICTURE	7
4	THE PHYSICAL FORM OF SOLID CATALYSTS: SURFACE AREA, SUPPORTS AND PROMOTERS	9
5	REQUIREMENTS OF A SOLID CATALYST	13
	5.1 Activity	14
	5.2 Selectivity	14
	5.3 Stability	15
	5.4 Summary of Sections 2–5	15
6	ADSORPTION	16
	6.1 Why do molecules adsorb at solid surfaces?	17
	6.2 Physical and chemical adsorption	17
	6.3 Chemical adsorption: a closer look	20
	6.4 Summary of Section 6	29
7	CHEMICAL ADSORPTION AND CATALYSIS	29
	7.1 A classification of solid catalysts	30
	7.2 Metals	31
	7.3 Metal oxides	34
	7.4 Acids	37
	7.5 Zeolites	38
	7.6 Summary of Section 7	44
8	ADSORPTION ISOTHERMS	44
	8.1 The Langmuir adsorption isotherm	45
	8.2 Other adsorption isotherms	50
	8.3 The determination of surface area	51
	8.4 Summary of Section 8	54
9	KINETICS OF SOLID-CATALYSED REACTIONS	55
	9.1 Rate of reaction, order, and effect of temperature	56
	9.2 Kinetic mechanisms	58
	9.3 Summary of Section 9	66
	OBJECTIVES FOR BLOCK 5	68
	SAQ ANSWERS AND COMMENTS	70
	ACKNOWLEDGEMENTS	82

1 INTRODUCTION

Many of the products that support our everyday lives – fuels, fertilizers, construction materials, medicines and artificial fibres, to name but a few – involve heterogeneous catalytic processes at one or several stages in their manufacture. By heterogeneous catalysis we mean that the catalyst has a distinct and separate phase from that of the reactants, so that the catalytic reaction takes place at a boundary, or *interface*, between phases. In most cases of practical interest the catalyst is a solid and the reactants are in a fluid phase (gas, liquid or solution), so that under these conditions the catalytic reaction takes place at the *surface* of the solid. Table 1 (overleaf) summarizes just a few of the many important processes of the modern petroleum-refining and chemical industries that depend on solid catalysts. *You don't need to memorize the details in Table 1; we shall refer to it on several occasions throughout this Block.*

The subject of heterogeneous catalysis is very broad and embraces a wide field of physical, chemical and engineering sciences. For instance, at one extreme the theories of catalysis are expressed in the language of solid-state physics: at another the mechanistic arguments of organic chemistry are used to formulate possible catalytic reaction mechanisms: at yet another the design of a commercial catalytic reactor relies heavily on chemical engineering principles. In this Block our treatment must necessarily be selective: our main aim is to examine the underlying reasons for the inherent *catalytic activity* of a variety of solid surfaces.

It is true that heterogeneous catalysis is often regarded as a black art. Certainly the mode of action of many industrial catalysts is still far from understood (or is a secret!). Coupled with this there is a vast amount of perhaps significant, but empirical, fact associated with the subject, and the predictive theories of catalysis are at best limited. Our approach is not to become overburdened with detail, but rather to attempt to elucidate general principles and so remove some of the 'aura of magic' from the subject. Much of the discussion will be based on chemical facts and ideas with which you are already familiar. In many cases the account will not be complete; this is because heterogeneous catalysis is an area of chemistry that is under development, and much remains to be discovered. Indeed, this is part of the excitement of the subject.

The theme of heterogeneous catalysis will be continued in Block 6, but the emphasis is different. There we shall examine modern physical methods for studying surfaces and the processes that occur on them. The information these 'surface science' techniques can provide is of somewhat controversial value for a number of reasons; for instance, (i) they usually detect only stable species, which may or may not be important in catalytic reactions, (ii) they may seriously disturb the surface, and (iii) the conditions they require, which often involve atomically clean crystal surfaces in an ultra-high vacuum environment, are far removed from those used in industrial processes. Nonetheless, we shall see in Block 6 that the 'gap' between surface science and practical catalysis can be at least partially bridged. Topic Study 2 brings together the principles and methods described in both Blocks and applies them to a current research area in catalysis.

STUDY COMMENT Video band 5 takes a short look at zeolites, which are discussed in Section 7.5 of this Block. There are no other video or AV sequences associated with this Block. However, the general ideas and concepts of heterogeneous catalysis will be brought up again in video bands associated with Block 6 and Topic Study 2.

Table 1 A selection of heterogeneously catalysed reactions of industrial importance.

Reaction[a]	Catalyst
'Heavy' inorganic industry	
$SO_2 + \frac{1}{2}O_2 \longrightarrow SO_3$ (sulfuric acid production)	Vanadium(V) oxide plus potassium sulfate on silica
$2NH_3 + \frac{5}{2}O_2 \longrightarrow 2NO + 3H_2O$ (nitric acid production)	90% platinum/10% rhodium wire gauze
$N_2 + 3H_2 \longrightarrow 2NH_3$ (ammonia production)	Iron promoted with aluminium, potassium, calcium and magnesium oxides
Hydrogenation	
animal and vegetable oils + $H_2 \longrightarrow$ edible fats	Raney nickel, or nickel on support
olefin (alkene) + $H_2 \longrightarrow$ paraffin (alkane)	Nickel, palladium or platinum on support
Reactions of carbon monoxide and hydrogen (synthesis gas)	
$CO + 2H_2 \longrightarrow CH_3OH$ methanol	Copper on zinc oxide with alumina
$CO + H_2 \longrightarrow$ hydrocarbons (Fischer–Tropsch syntheses)	Iron or cobalt, on support, with promoters
Catalytic cracking	
naphtha[b] $\longrightarrow$ low molecular mass hydrocarbons for fuel	Solid acids; for example, amorphous silica–alumina, or zeolites
Hydrocarbon processing	
refinery, or catalytic reforming: hydrocarbons $\rightarrow$ high-grade petroleum; hydrocarbons $\rightarrow$ benzene, toluene and xylenes (collectively known as BTX)	Platinum, platinum/rhenium, platinum/iridium or platinum on acidified γ-alumina
steam reforming: CH_4 (or other hydrocarbons) + $H_2O \rightarrow CO + H_2$ (synthesis gas)	Nickel or copper on supports such as alumina, magnesia or mixtures thereof
Dehydrogenation	
$C_4H_8 \longrightarrow CH_2{=}CHCH{=}CH_2 + H_2$ butene → buta-1,3-diene	Chromium oxide/alumina
$C_6H_5C_2H_5 \longrightarrow C_6H_5CH{=}CH_2 + H_2$ ethylbenzene → styrene	Iron(III) oxide promoted with chromium oxide and potassium carbonate
Selective oxidation	
$CH_2{=}CH_2 + \frac{1}{2}O_2 \longrightarrow (CH_2)_2O$ ethene → ethylene oxide (oxirane)	Supported silver
$C_4H_{10} + O_2 \longrightarrow CH_2{=}CHCH{=}CH_2 + 2H_2O$ butane → buta-1,3-diene	Ferrite spinels (for example, $MnFe_2O_4$) or bismuth molybdates supported on silica
$CH_2{=}CHCH_3 + NH_3 + \frac{3}{2}O_2 \longrightarrow CH_2{=}CHCN + 3H_2O$ propene → vinyl nitrile (acrylonitrile)	Complex metal molybdates or multimetallic oxide compositions supported on silica
$CH_2{=}CHCH_3 + O_2 \longrightarrow CH_2{=}CHCHO + H_2O$ propene → propenal (acrolein)	Bismuth molybdates supported on silica
$CH_2{=}CH_2 + \frac{1}{2}O_2 + CH_3COOH \longrightarrow CH_3COOCH{=}CH_2 + H_2O$ ethene → vinyl acetate	Palladium on suitable support
Control of pollution	
Removal of unburnt hydrocarbons, CO, and nitrogen oxides from motor vehicle exhausts. Complete oxidation of hydrocarbons, oxidation of CO.	Platinum, platinum/palladium or platinum/rhodium supported on ceria doped alumina

[a] Because of the complexity of many heterogeneously catalysed reactions, we have used the arrow sign rather than the equality sign in writing overall chemical reactions throughout this Table.

[b] The composition of naphtha depends very much on the source of the crude oil. It may be regarded as the fraction that distills from crude oil in the temperature range from 340 to 460 K. It contains in excess of 400 hydrocarbons with between four and ten carbon atoms each (C_4–C_{10} in common notation).

2 A DEFINITION

The definition of a catalyst given in Block 4 also applies, *as it must*, to a solid catalyst. For the purpose of this Block the most useful form of the definition is:

> A catalyst is a substance that increases the rate of a chemical reaction without itself being consumed: it does not alter the position of thermodynamic equilibrium.

Implicit in this definition is that the amount of matter changed in a reaction is many times greater than that of the catalyst present.

Popular definitions of a catalyst are often along the lines that 'it increases the rate of a chemical reaction without itself being *affected*'. This is misleading. In practice, solid catalysts are often found to change in structure owing to their participation in a reaction. For instance, some metal catalysts frequently change in surface roughness or crystal structure during use: an example is given in Figure 1.

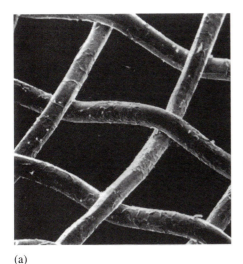
(a)

(b)

Figure 1 Scanning electron micrographs of the platinum alloy gauze (90% platinum/10% rhodium) used in the industrial oxidation of ammonia. The gauze wire is 0.075 mm in diameter. (a) Fresh gauze. (b) The same section after one-half the normal lifetime of gauze installed in a nitric acid plant. Detailed investigation reveals that reconstruction of the active catalyst surface probably occurs throughout the working life of the catalyst.

3 A SIMPLIFIED POTENTIAL ENERGY PICTURE

In order to introduce some of the key issues in heterogeneous catalysis, it is useful to construct a potential energy picture. To do this we can divide the *total* process of heterogeneous catalysis into the following five steps.

1   Diffusion of reactant molecules to the solid surface: this is referred to as **mass transfer** or **mass transport**. If the catalyst is porous – which, as we shall see, is often the case with industrial catalysts – then two stages are involved: reactant molecules are first transported to the outside surface of the catalyst and then pass into the catalyst through the pore structure.

2   Attachment of *at least one* of the reactants to the solid surface: this is called **adsorption**.

3   A chemical reaction *on the surface* of the catalyst involving one or more adsorbed species. (Notice that this step does not exclude a surface reaction between an adsorbed reactant and a reactant in the gas phase.)

4   Detachment of the products from the catalyst surface: this is called **desorption**.

5   Diffusion of desorbed products away from the catalyst. Two stages – the reverse of those described in step 1 – are involved if the catalyst is porous.

■   Which of these steps do you think require an energy of activation?

□   In principle, they all *may* require an energy of activation and thus, depending on the reaction, any one may be rate-limiting.

Steps 2, 3 and 4 are chemical in nature: jointly they can be considered to constitute the *chemical processes* involved in a solid-catalysed reaction. Steps 1 and 5 are *physical processes* involved with mass transport. If either of these is rate-determining, then the overall catalytic reaction is said to be *diffusion-limited*.

Figure 2 shows a schematic potential energy profile for an *exothermic*, heterogeneously catalysed gas-phase reaction that is *not* diffusion-limited.

■   In Figure 2, what do the labels A and B for step 2 correspond to?

□   Step 2 is the adsorption process; notice that its potential energy profile is very similar to that for an elementary chemical reaction, which we discussed in Block 2. Thus, by analogy, label A represents the **enthalpy of adsorption**, ΔH_{ad}, where

$$\Delta H_{ad} = H(\text{adsorbed state}) - H(\text{reactants}) \tag{1}$$

Since $H(\text{adsorbed state}) < H(\text{reactants})$, then $\Delta H_{ad} < 0$, and so the adsorption process is exothermic. Label B corresponds to the activation energy for the adsorption of reactants.

■   What does the label C for step 3 correspond to?

□   Step 3 is the *surface* chemical reaction, and so label C represents the activation energy for this process. The maximum in the curve corresponds to the transition state for the surface reaction.

■   What does the label D for step 4 correspond to?

□   Step 4 is the desorption process and so label D corresponds to the activation energy for this process.

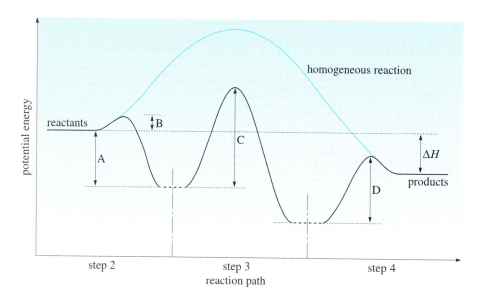

Figure 2   Schematic potential energy profiles for an exothermic gas-phase reaction occurring in the presence (black line) and absence (blue line) of a solid catalyst under the same experimental conditions. The step numbers refer to those described in the text.

Generally speaking, the ability of a solid catalyst to increase a reaction rate can be ascribed to the fact that it provides an alternative pathway to chemical reaction that is energetically more favourable than the one that would occur entirely in the gas phase (the homogeneous reaction) – as shown schematically in Figure 2. This in turn means that the overall activation energy for reaction is considerably *less* than that for the corresponding homogeneous reaction: indeed, this is the essence of heterogeneous catalysis.

Of course, it must be admitted that the potential energy picture in Figure 2 is very much simplified. But it does allow us to highlight the action of a solid catalyst in a very concise way:

> The surface of a solid catalyst is important because it is there that the reaction takes place. Given the 'right' catalyst surface, adsorption of reactant molecules can take place and, while these molecules are temporarily restrained, chemical reaction occurs to give product molecules, which are then desorbed.

Clearly, an understanding of both adsorption and the chemistry of surface reactions is vital in appreciating how a solid catalyst works.

There is one final, but important, point to make concerning solid catalysts before leaving this Section. *All* solid catalysts are themselves non-uniform, or heterogeneous, in the sense that on the atomic scale their chemical and physical properties *vary with location on the surface*. Even in a pure metal catalyst, the atoms at specific locations such as lattice defects, and edges and corners of crystallites (very small crystals), are different from atoms on the crystal faces. In addition – and more importantly as you will see in Block 6 – the surface atoms on different crystal faces have different properties.

The idea that surfaces are heterogeneous has long been known. In 1925, H.S. Taylor introduced the concept that reactions take place *only* at specific locations on a catalyst surface; these locations are called **active sites** (or **active centres**). The idea of active sites is still current today although, despite the introduction of sophisticated surface science techniques, their exact identity and structure are often elusive. In some cases, an active site may be a small group or cluster of atoms: in others, it may actually be a species attached to the surface of the catalyst.

4 THE PHYSICAL FORM OF SOLID CATALYSTS:
SURFACE AREA, SUPPORTS AND PROMOTERS

Heterogeneous catalysis requires the presence of a surface. It thus follows that the rate of a catalysed reaction will, under most conditions, depend on the **surface area** of the catalyst exposed to the reactants: for a given mass of catalyst a larger surface area will usually result in a faster reaction. Thus, apart from its chemical composition, the physical form in which a catalyst is manufactured is a major factor in determining its activity. This is illustrated in Figure 3 for the **water-gas shift reaction**:

$$CO(g) + H_2O(g) = CO_2(g) + H_2(g) \quad (2)$$

which is catalysed by a multicomponent catalyst, $Cu-ZnO-Al_2O_3$. The plot shows that the activity of the catalyst, which may be taken as just a measure of the rate of reaction expressed in arbitrary units, increases approximately linearly as the exposed area of the active copper species in the catalyst increases. (All the data are for the special case that there are no diffusion limitations, that is the overall process is

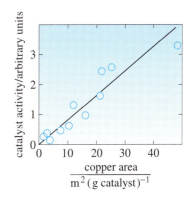

Figure 3 Relationship between the surface area of a catalyst and catalytic activity for reaction 2.

reaction-rate limited, the kinetics being determined solely by the rate of the surface reaction(s). Although the corresponding industrial process is certainly diffusion-limited, experiment shows that the catalyst activity still increases, roughly as the square root of the exposed copper surface area.)

In general, surface areas of commercial catalysts range from about 10 to 1 000 m² per gram of catalyst. A simple classification is often used: 'low' refers to surface areas up to $10\,m^2\,g^{-1}$, and 'very high' to surface areas of $200\,m^2\,g^{-1}$ and greater, such as in special aluminas. It is important to appreciate that these areas are enormous. For just *one gram* of catalyst, the effective surface area can be comparable with that of a tennis court, and for the 100 tonnes or so of catalyst used in a typical industrial reactor it can be of the order of 2 000 square miles: that is roughly one third of the area of Yorkshire. (We shall see how to measure surface areas in Section 8.3.)

In addition to the surface area, there are other important physical factors to be taken into account in designing and making a catalyst. For instance, the catalyst must be stable at the high operating temperatures that are often necessary to give commercially acceptable reaction rates, and it must also be mechanically robust so that it does not disintegrate in the harsh environment of the reactor. It is these considerations, and others, that account in large part for the multicomponent composition of many of the catalysts in Table 1. Indeed, the so-called 'structural engineering' of catalysts is an involved subject and we have space here only to scratch the surface. However, it is a science. To quote one expert:

> In the recent past several generalities have become discernible in catalyst preparation and it may now be claimed, without arousing excessive merriment, that a catalyst preparation science is being built.
>
> (S.P.S. Andrew, 'The black art of designing and making catalysts', in *Chem. Tech.*, American Chemical Society, 1979, vol. 9, pp. 180–184)

A common way to achieve a high surface area for a metal or metal oxide that is catalytically active is to prepare it in the form of very small crystallites.

SAQ 1 The size of these crystallites can be related to the total surface area of a given mass of catalyst if it is assumed that the crystallites are all uniform and spherical in shape. Using this approximation show that:

$$\text{total surface area of crystallites} = 6M/\rho d \tag{3}$$

where M is the total mass of the catalyst in the form of crystallites, ρ is its density and d is the diameter of a single crystallite. [*Hint* Remember that the volume of a sphere of radius r is $\frac{4}{3}\pi r^3$ and its surface area is $4\pi r^2$. You will need to write an expression for the mass of a *single* crystallite, and hence determine the *number* of crystallites in a mass M.]

If the catalyst is platinum metal ($\rho = 21.45 \times 10^6\,g\,m^{-3}$) and the spherical crystallites are of 5 nm diameter, what is the total surface area available from 1 g of platinum?

The calculation in the answer to SAQ 1 shows that very small crystallites are needed if the surface area exposed per gram of platinum is to be maximized: the smaller the crystallites, the larger the total surface area. For such a costly metal as platinum this is a very significant consideration. But it also raises an extremely important general problem. Most metals or metal oxides will **sinter** readily if they are in the form of very small crystals (say, less than 50 nm diameter), especially if they are at temperatures above about one half their melting temperature (known as the *Tammann temperature*). The effect of sintering is to cause crystallites to coalesce and so grow in size: this has the undesirable effect of reducing the surface area. Thus the prevention of sintering is a major factor to take into account when designing a catalyst. It is, however, a factor that is difficult to treat in isolation and so generalizations must be treated with some caution.

For metal (and other) catalysts, sintering can be prevented by dispersing them in the form of very small particles on, or within, an inert and usually porous material of high total surface area: this material is called a **support** or **carrier**. One way to do this is by **impregnation**: porous lumps, or pre-formed pellets, of the support are soaked in a solution containing a salt of the metal and then they are dried. The composite is then *activated*, that is converted into its active form, through physical and chemical changes. Typically, this involves heating to cause decomposition of the metallic salt, followed by reduction to give the metallic catalyst. Figures 4 and 5 show electron micrographs, obtained using transmission electron microscopy, of two supported metal catalysts. Typically the active ingredient constitutes about 0.1–20% by mass of the composite.

SAQ 2 A palladium/alumina catalyst was prepared by impregnation of 10 g of the alumina support with 5 g of a palladium precursor solution with an assay (by mass) of 8.13% palladium. Assuming that only 90% of the precursor is taken up and that, after activation, only metallic palladium remains on the surface, what is the palladium content of the catalyst (in mass %)? [*Hint* Assay gives the percentage by mass of the metal in the solution, so it is not necessary to know the molecular mass of the precursor.]

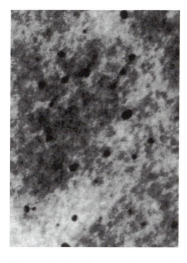

Figure 4 An electron micrograph of a 'slice' through a silica-supported palladium/gold alloy catalyst (4% gold, 1% palladium by mass). The black dots are particles of the alloy.

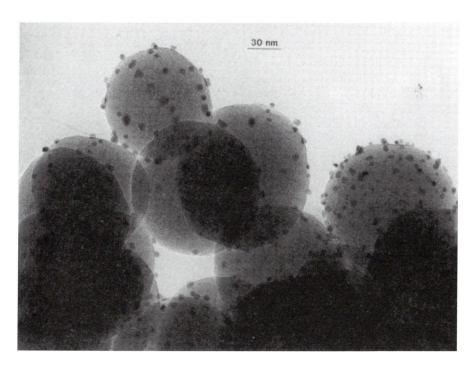

Figure 5 An electron micrograph of a supported metal catalyst, rhodium/silica. The metal crystallites are present on the surfaces of particles of silica.

The most widely used support material is alumina, although silica and activated carbon are sometimes employed. Alumina can be prepared in a number of crystallographic forms. Gamma-alumina, γ-Al_2O_3 (also called activated alumina), is a porous material of very high surface area – typically greater than 200 $m^2\,g^{-1}$ – and is relatively stable over the temperature range of interest for most catalytic reactions. Alpha-alumina, α-Al_2O_3, is chemically inert, has a low surface area and finds applications in catalysts that are to be used under arduous conditions.

Although a digression from our main theme, it is worth noting that support materials are not always chemically inert: indeed, as you will see later in this Block, γ-Al_2O_3 has catalytic activity in its own right. In **dual-function catalysts** – the best known examples being those used for reforming reactions (cf. Table 1) – the acidified γ-Al_2O_3 both acts as a support for the metal or alloy and contributes *chemically* to the overall process.

Another means of preventing sintering in catalysts in which the active species is a metal with a relatively low melting temperature is to include in the catalyst non-sintering inorganic materials, described variously as **stabilizers**, **structural promoters**, or **textural promoters**. The oxides of aluminium, chromium and magnesium are very commonly used. When dispersed among the metal crystallites, these refractory materials serve to separate the metal crystallites and so prevent them joining together. In this way the active metal ingredient can constitute in some cases

75–95% by mass of the total catalyst. A good example of structural promotion is provided by the iron catalyst used in the production of ammonia. In its active form it is typically composed mainly of metallic iron in the form of very small crystallites, of between 20 and 40 nm overall dimensions, separated by a mixture of structural promoters including alumina, silica, magnesia (MgO) and calcium oxide. Such a catalyst is called **multiply-promoted** and is highly porous.

Note that not all promoters have a structural role. **Chemical or electronic promoters**, as they are justly called, are thought to change the chemical composition of a catalyst with the result that the intrinsic activity of the surface is increased; however, the mechanism of their action is not always clear. Again the ammonia catalyst provides a classic example. The addition of only about 0.8% by mass of potassium oxide, K_2O, has a marked effect on the overall activity of the iron catalyst.

Figure 6a shows the coarse particle or pellet form in which most catalysts are prepared for use in industrial reactors. Pellets are often shaped so as to aid the flow of gas through a reactor and are typically about the size of aspirin tablets. A further important, and quite different, form of catalyst is the so-called monolithic structure shown in Figure 6b, which typically consists of a block of ceramic material through which runs a honeycomb of parallel channels. The active phase is dispersed through the inside of the monolith in a thin layer of porous alumina. This type of catalyst is finding application particularly in air-pollution control.

SAQ 3 One way to classify industrial catalysts is in terms of their *physical* form. The following categories are often referred to: (i) unsupported metals or alloys; (ii) very high surface area materials; (iii) supported dispersions: binary systems; (iv) supported, or unsupported, multicomponent systems. Look at Table 1 (Section 1) and see if you can identify examples of each of these categories.

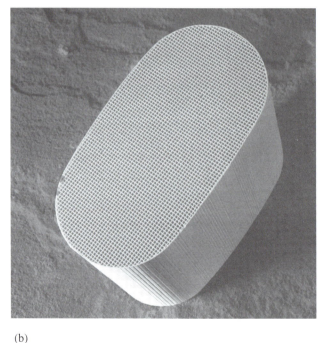

(a) (b)

Figure 6 (a) A selection of catalysts used by Imperial Chemical Industries Katalco. (b) A monolith.

5 REQUIREMENTS OF A SOLID CATALYST

Among the various factors that determine the suitability of a catalyst for an industrial reaction the following three predominate: *activity, selectivity* and *stability*. We consider each of these separately below.

Before any thought is given to the selection of a catalyst, it is always *essential* to establish that a proposed reaction is thermodynamically feasible. For instance, suppose we wish to manufacture hydrogen cyanide, HCN, which is an important chemical feedstock for the production of nylon and various transparent plastics. A possible reaction is as follows:

$$CH_4(g) + NH_3(g) = HCN(g) + 3H_2(g) \tag{4}$$

■ Is reaction 4 thermodynamically favourable under standard conditions at 298.15 K?

▫ A reaction is thermodynamically favourable if $\Delta G_m^\ominus$ is negative. Using information from the S342 *Data Book*:

$$\Delta G_m^\ominus = \Delta G_f^\ominus (HCN, g) + 3 \Delta G_f^\ominus (H_2, g) - \Delta G_f^\ominus (CH_4, g) - \Delta G_f^\ominus (NH_3, g)$$

$$= (124.7 + 0 - (-50.7) - (-16.4)) \text{ kJ mol}^{-1}$$

$$= 191.8 \text{ kJ mol}^{-1}: \text{a large positive number}$$

Obviously, the reaction is not at all favourable at 298.15 K, and a catalyst would be of no use under these conditions because the equilibrium yield of the products would be negligible.

> **STUDY COMMENT** The following SAQ draws on the discussion of the thermodynamic 'limits of operation' in Block 1. It would be a good idea to have a go at it (and check our answer) at this point, before moving on.

SAQ 4 (revision) The equilibrium yield of hydrogen cyanide from reaction 4, which is highly endothermic, will be significant only at high temperatures. (Remember that in this Course, as explained in Block 1, we define the *equilibrium yield* of the *desired product* in a gaseous reaction as $p(\text{product})/p_{\text{tot}}$, where p_{tot} is the total pressure of the equilibrium mixture.)

(a) Determine the temperature at which the equilibrium yield of hydrogen cyanide is 20%. (Assume that the overall pressure is 1 bar and that the reactants are mixed in stoichiometric proportions.)

(b) State carefully any assumptions that you make in your calculations.

The answer to SAQ 4 stresses a very significant point. It is thermodynamic considerations that dictate, in the first instance, the conditions – pressure, and particularly temperature – under which a catalyst must operate if reasonable conversion into products is to be achieved. Of course, industrial reactions are not necessarily allowed to go to final thermodynamic equilibrium, and unused reactant gases are often recycled so that several passes are made through the catalytic reactor. This can be extremely important when the theoretical equilibrium yield under the prevailing conditions is low, as for example in ammonia synthesis, where it is about 20%. For a given process there is thus an incentive to find a catalyst that will give good conversion into products during a certain **contact time** with the catalyst under mild reaction conditions. Indeed, the economic viability of a process may often depend on this.

5.1 Activity

As its name implies, and as we have hinted at already, the **activity** of a catalyst refers to its ability to transform reactant(s) into product(s): it is commonly quoted in terms of unit mass or unit volume of catalyst under specified conditions. In quantitative terms it can be measured as a rate of reaction, and we shall say more about this later, in Section 9. As in the example in Figure 3 (Section 4), in many circumstances the activity is simply quoted as a value relative to some standard: alternatively, some subjective or qualitative estimate may be used. It should also be remembered that the intrinsic activity of a catalyst can be reduced by sintering if the temperature of a process is raised.

5.2 Selectivity

The **selectivity** of a catalyst relates to its ability to enhance the rate of just one of several thermodynamically feasible reactions. This can be illustrated by looking at a well-known example.

In the gas phase, ethanol, C_2H_5OH, could possibly decompose either by the loss of hydrogen – that is a *dehydrogenation* reaction – to give ethanal, CH_3CHO:

$$C_2H_5OH(g) = CH_3CHO(g) + H_2(g) \tag{5}$$

or by the loss of water – that is a *dehydration* reaction – to give ethene, C_2H_4:

$$C_2H_5OH(g) = C_2H_4(g) + H_2O(g) \tag{6}$$

SAQ 5 (revision) Calculate values of $\Delta H_m^\ominus$ and $\Delta S_m^\ominus$ at 298.15 K for reactions 5 and 6. Hence estimate values of $\Delta G_m^\ominus$ and the standard equilibrium constant, $K^\ominus$, at 600 K and 1 000 K, for both reactions. What do you conclude from your answers?

The answer to SAQ 5 shows that over the range 600–1 000 K reaction 6 is thermodynamically more favourable than reaction 5. However, it is possible to *selectively* enhance the rate of either one of these reactions by judicious choice of the right catalyst. In particular, around 600 K, a copper catalyst promotes the less favourable dehydrogenation reaction. An alumina catalyst favours dehydration.

The decomposition of ethanol is an example of a catalysed reaction in which there are alternative, thermodynamically feasible, reaction paths leading to different products. However, there are also numerous catalysed reactions that involve successive reaction steps leading to one set of end-products, *but* in which thermodynamically favourable *intermediate* species may be isolated. The industrial synthesis of propenal, or acrolein, $CH_2=CHCHO$, via the partial oxidation of propene, $CH_2=CHCH_3$, provides an example:

$$CH_2=CHCH_3(g) + O_2(g) \longrightarrow CH_2=CHCHO(g) + H_2O(g) + \text{other products} \tag{7}$$

A number of other partially oxidized products (e.g. CH_3CHO and $CH_2=CHCOOH$) could result from this reaction, but a particular type of bismuth molybdate catalyst* is highly selective towards acrolein under the conditions used for the process. The acrolein is, however, an intermediate product and can be isolated only if the contact time between the catalyst and reactants is relatively short (typically, for the conditions of the industrial process, a second or so). Complete oxidation to carbon dioxide and water would certainly occur if the reactants were allowed to remain in contact with the catalyst indefinitely:

$$CH_2=CHCH_3(g) + \tfrac{9}{2}O_2(g) = 3CO_2(g) + 3H_2O(g) \tag{8}$$

---

* Bismuth molybdate catalysts have complex structures based on a mixture of MoO_3 and Bi_2O_3 in various proportions. This is discussed further in Section 7.3.

Selectivity is determined by the chemical nature of the catalyst, but it also depends very much on the reaction conditions – temperature, pressure, reactant composition, and, as we have seen above, contact time. On balance, selectivity is often the main factor to be considered in choosing an industrial catalyst (low conversions can be increased by recycling unreacted reactants). To summarize: the industrialist more often than not requires a catalyst that will enhance *only* the formation of the desired product with a minimum of side-reactions.

5.3 Stability

A catalyst must have sufficient **stability** to enable it to retain its activity and selectivity over a reasonable length of time. How long is a matter of judgement: some industrial processes have catalysts that last for several years.

A catalyst may deteriorate during use for a number of reasons. One common cause is sintering: others can be roughly grouped under the headings of (a) poisoning and (b) fouling.

Poisoning A catalyst poison is an impurity present in the reaction mixture, which severely reduces catalyst activity: notorious examples are sulfur and chlorine, or compounds incorporating these elements, which readily attach to, and hence prevent access to, active sites on the catalyst surface. In some cases poisons can also induce sintering, for example the effect of halogens on copper catalysts. Poisoning may be reversible in favourable cases, so that with suitable treatment full catalyst activity can be regenerated. In addition, selective poisoning can be used to improve selectivity, for example in the Lindlar catalyst (lead-poisoned palladium on $CaCO_3$) used for hydrogenation reactions.

Fouling This refers to a physical blockage, by dust or carbonaceous deposits, which prevents reactants from reaching the catalyst surface.

These effects can be prevented to some extent by a third group of promoters, **poison-resistant promoters**, which protect the active phase against poisoning or fouling, either by impurities in the reactants or by substances generated during side-reactions. One example of this is the use of platinum/rhenium catalysts for petroleum reforming, where one of the effects of rhenium is to protect the platinum against carbon deposition, which can foul the active surface.

> **STUDY COMMENT** The following SAQ gives you a chance to summarize the key points in this subsection.

SAQ 6 The industrial production of methanol is based on the hydrogenation of carbon monoxide (equation 9)

$$CO(g) + 2H_2(g) = CH_3OH(g) \tag{9}$$

Referring back to the information in Exercise 1 at the end of Block 1, list the properties you would look for in selecting a catalyst suitable for methanol synthesis.

5.4 Summary of Sections 2–5

1 A solid catalyst is a substance that increases the rate of a chemical reaction without itself being consumed: it does not alter the position of thermodynamic equilibrium. The physical state of a solid catalyst may change during use.

2 The overall process of heterogeneous catalysis can be broken down into five steps, each of which may require an energy of activation. The steps are: (1) mass transfer, or transport, of reactants to the active sites of the catalyst; (2) chemical adsorption of at least one reactant; (3) a surface chemical reaction involving one or more adsorbed species; (4) desorption of products; and (5) diffusion of desorbed products away from the catalytic surface. Steps 1 and 5 are physical processes; steps 2, 3 and 4 are chemical processes.

3   In general terms, a solid catalyst increases a reaction rate because of its ability to provide an alternative, and more energetically favourable, pathway to reaction.

4   All solid catalysts are themselves heterogeneous in that their surfaces are non-uniform on the atomic scale. Catalytic reactions are thought to take place at active sites (or centres).

5   Under most conditions the rate of a heterogeneously catalysed reaction depends, among other factors, on the surface area of the active component of the catalyst that is exposed to the reactants.

6   Industrial catalysts have surface areas typically in the range from 10 to 1 000 $m^2\,g^{-1}$ and are often prepared in a porous form.

7   Sintering is the process by which very small crystals of metals or metal oxides coalesce to form larger particles: the rate of sintering depends on both the temperature and the size of the crystals.

8   A catalyst support or carrier, for example α- or γ-alumina, is used to increase and maintain the surface area of an active catalyst: it is not always chemically inert.

9   A promoter is a substance that, when added to a catalyst in small amounts, prevents sintering (structural or textural promoter) *or* enhances the activity (chemical or electronic promoter) *or* protects against poisoning or fouling (poison-resistant promoter).

10 The requirements of a catalyst for an industrial reaction, apart from cost, are determined by three main factors: activity, selectivity and stability. Activity relates to the rate of a catalysed reaction, selectivity to the ability to catalyse preferentially one of several reactions, and stability to the lifetime, or ease of regeneration, of a catalyst. The choice of a particular catalyst must be made within the limitations imposed by thermodynamic considerations.

6 ADSORPTION

So far, we have discussed some of the practical aspects of heterogeneous catalysis. In this Section we turn our attention to the more fundamental issue of how a solid catalyst works. The best starting point is to examine in some detail the underlying chemistry of heterogeneous catalysis, and this in turn means that we must learn more about the key process of adsorption.

It is worthwhile at the outset to be clear about the nomenclature we shall use. As already stated in Section 3, *adsorption* describes the process by which atoms or molecules are attached to a solid surface: it is purely a surface effect. The material that adsorbs on the surface is the **adsorbate** and the underlying solid is the **adsorbent**, although to avoid possible confusion we shall always use the alternative term, **substrate**. In the case of porous solids, gas molecules can diffuse into their interiors through the pore structure, but the molecules still *adsorb* on the walls of the pores*.

---

* If a chemical species penetrates the lattice of a solid by a process similar to dissolution, this is termed *absorption*. If a process involves both adsorption and absorption, or there is uncertainty as to its exact nature, it is called *sorption*.

6.1 Why do molecules adsorb at solid surfaces?

To begin, let us answer this question in very simple terms.

- Can you recall the three groups into which most solids can be divided depending on their bonding?
- The three groups are *ionic* solids, *molecular* solids, and *metallic and giant molecular* solids.

If we put enzymes to one side as a special case, in broad terms it turns out that it is ionic and metallic materials that can act as solid catalysts.

- What are the main features of the bonding in ionic and metallic solids?
- In an ideal ionic solid it is assumed that the electrons are transferred from one atom to another to give cations and anions, and that these ions are held together by strong electrostatic forces. The bonding in metallic solids can be pictured in a simple way in terms of positive ions held together by valence electrons, which are delocalized throughout the crystal.

Figures 7a and b depict, in an idealized way, the bonding in metallic and ionic solids, respectively.

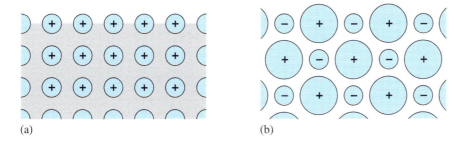

Figure 7 An idealized representation of the bonding in (a) a metal and (b) a simple ionic solid.

With reference to Figure 7, can you suggest what makes the surfaces of metals and ionic solids so special?

They, and indeed the surfaces of *all* solids, are special because the surface atoms are in the unique position of not having a full complement of nearest neighbours as they would in the bulk: consequently the atoms exert a net attractive force on any species approaching the surface and this favours adsorption. In the case of ionic solids this is mainly electrostatic in nature, whereas in the case of metals it can be thought of as arising from the free valencies of the surface metal atoms. Of course, this picture is very simplified – for instance the electronic structure of a metal surface is highly complex – but, nevertheless, in essence it provides a way of seeing why molecules adsorb at surfaces. Our next step must be to refine these ideas.

6.2 Physical and chemical adsorption

Adsorption phenomena have for many years been recognized to be of two types: *physical adsorption* and *chemical adsorption*; the latter is often referred to as *chemisorption*.

6.2.1 Physical adsorption

In **physical adsorption** the forces of attraction between the adsorbate and solid surface are intermolecular, that is they are similar in nature to the attractive, cohesive forces that are responsible for the condensation of a vapour to a liquid. It is thus not surprising that the strength of physical adsorption depends on the *physical* properties of the adsorbate species. The amount of a gas physically adsorbed always increases with decreasing temperature, and is most readily determined at temperatures close to the normal boiling (or condensation) temperature of the adsorbate. Thus physical adsorption of inert gases and non-polar molecules such as nitrogen and hydrogen is weak, and detectable only at very low temperatures (lower than, say, 150 K). On the other hand, for polar molecules with permanent dipoles, such as water and benzene, physical adsorption is relatively strong and readily detectable around 373 K.

The process of physical adsorption can usefully be represented by a potential energy curve. The direct calculation of such a curve for the physical adsorption of a molecule, or even more simply an atom, on the surface of a solid is extremely difficult, since in principle the interaction energy with *all* surface atoms must be included. Nonetheless, good approximations can be obtained for the general shape of such a curve. For example, Figure 8 depicts in a schematic way the interaction of a hydrogen molecule with surface atoms of a metal, which for convenience we have labelled M; the zero of energy has arbitrarily been chosen to correspond to the situation in which the hydrogen molecule is at an infinite distance from the surface. To get above the axis of zero energy, energy has to be supplied: if the system falls below it, energy is liberated.

In Figure 8, the quantity r_p represents a distance and the quantity E_p an energy. What is the physical interpretation of each of these quantities?

The quantity r_p is the *equilibrium* distance of the hydrogen molecule from the surface of the metal. Typically, calculations indicate that it lies in the range 0.3–0.4 nm, a distance considerably longer than 'normal' chemical bond lengths. The quantity E_p is the amount of energy *released* when the hydrogen molecule is physically adsorbed. To a good approximation, it is reasonable to equate this energy difference with the **enthalpy change for physical adsorption**, ΔH_{pa}. It is important to note that physical adsorption is *always* exothermic; in this case:

$$\Delta H_{pa} = H(\text{H}_2 \text{ physically adsorbed on surface}) - H(\text{surface} + \text{free H}_2 \text{ molecules}) < 0 \quad (10)$$

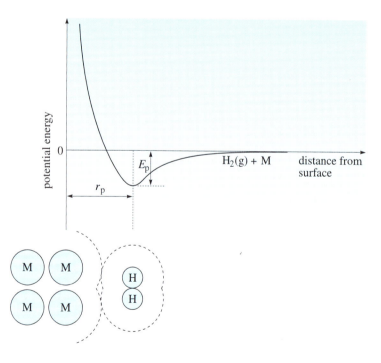

Figure 8 Schematic representation of the potential energy curve for the physical adsorption of a hydrogen molecule on the surface of a metal, M. A pictorial view of the adsorbed state is also shown, but it must be emphasized that this is drawn purely to indicate the general nature of physical adsorption rather than to indicate the exact details of the weak bonding.

The *magnitude* of ΔH_{pa} in a given system is usually less than about 40 kJ mol^{-1}; and, indeed, it is often comparable in magnitude with the enthalpy of condensation of the gaseous adsorbate on the surface of its own liquid. Hence, on a chemical scale, the energies involved in physical adsorption are relatively small and are certainly insufficient to lead to the breaking of chemical bonds. Thus on physical adsorption a molecule will *retain* its chemical identity, although there may be small changes in geometry owing to the proximity of the surface. Since the forces involved are relatively non-specific, physical adsorption will occur at all surfaces: clearly, it cannot be the key step in the mechanism of heterogeneous catalysis. But it should not be dismissed out of hand. As you will see in due course, physical adsorption does have an important role to play in the overall process of heterogeneous catalysis. In addition, the phenomenon lies at the heart of a widely used method for determining the surface area of solids.

SAQ 7 Would you expect an activation energy to be required for (a) physical adsorption, and (b) desorption of a physically adsorbed species? Illustrate your answer to both parts of the question by reference to the schematic potential energy profile in Figure 8.

6.2.2 Chemical adsorption

Chemical adsorption is very specific and can occur only if the adsorbate is capable of forming a *chemical bond* with the surface. It can be thought of as a chemical reaction, because it involves fairly substantial rearrangement of electrons between the adsorbate and substrate. As you might guess, the energy changes are far greater than in physical adsorption, and **enthalpy changes for chemical adsorption**, ΔH_{ca}, are commonly found to have values in the range from −40 to −400 kJ mol^{-1}.

It is sometimes useful to distinguish between two types of chemical adsorption. In **dissociative chemisorption** a molecule breaks into two or more fragments, all of which adsorb (temporarily) on the surface. For example, it is well established that hydrogen, H$_2$, generally dissociates into hydrogen atoms upon chemical adsorption on a metal surface. We can represent this 'reaction' as:

$$H_2(g) + 2* \longrightarrow \underset{*}{\overset{H}{|}} + \underset{*}{\overset{H}{|}} \tag{11}$$

where each asterisk denotes an adsorption site on the metal surface. Another example is the dissociative adsorption of propene, CH$_2$=CHCH$_3$, on certain metal oxides:

$$H_2C{=}CHCH_3(g) + 2* \longrightarrow H_2C\text{---}\underset{*}{\overset{\overset{H}{|}}{C}}\text{---}CH_2 + \underset{*}{\overset{H}{|}} \tag{12}$$

to form a π-bonded intermediate (or symmetric allyl radical) and a hydrogen atom.

> **STUDY COMMENT** In reaction 12 the bonding in the intermediate is depicted in a schematic way. Types of chemical bonding are discussed in the Second Level Inorganic Course, which you should refer to if you find difficulty with any of the descriptions of bonding in this or later Sections.

In **associative chemisorption** (also termed *non-dissociative chemisorption*) a molecule adsorbs without fragmentation. For example, the adsorption of carbon monoxide is associative on certain metal surfaces and can be represented schematically as:

$$CO(g) + * \longrightarrow \underset{*}{\overset{O}{\underset{\|}{\overset{\|}{C}}}} \tag{13}$$

An interesting case is the adsorption of ethene, C_2H_4, on a metal surface, because one possible surface intermediate is a di-σ-bonded species:

$$H_2C=CH_2(g) + 2* \longrightarrow \begin{array}{c} H_2C-CH_2 \\ | \quad\; | \\ * \quad\; * \end{array} \qquad (14)$$

Even though the adsorption is classified as associative on the basis that no fragmentation occurs, it is important to notice that two *adjacent* surface sites are required for this type of adsorption.

SAQ 8 Can you suggest how methane gas, $CH_4(g)$, may chemically adsorb on a metal surface?

The nature of chemical adsorption means that *at most* only a single layer of adsorbate molecules can be formed at a surface: that is, it is restricted to **monolayer coverage**. By contrast, in physical adsorption **multilayer formation** is quite feasible because the processes involved are very much akin to those of condensation. Physical adsorption and chemical adsorption can occur together, but any adsorbed layers beyond the first are physically adsorbed.

It should be clear that the major feature of chemical adsorption is that the *chemical character* of a molecule is changed: indeed it may even be torn apart! This then is the type of interaction that is important in heterogeneous catalysis: it attaches molecules, or their fragments, to a surface in such a way that they are 'ready' to take part in a surface reaction. In fact, there is little doubt that the chemical adsorption of *at least one* reactant is a key step in *all* solid-catalysed reactions. So a deeper understanding of chemical adsorption phenomena on catalysts should provide insight into the mechanism of their action.

6.3 Chemical adsorption: a closer look

We shall restrict our discussion here to metal substrates and gaseous adsorbates. Apart from the fact that many metals are themselves catalysts, the main reason for this restriction is that adsorption on metals has been extensively studied and it is possible to extract from the large body of available information several features that are of direct, and general, relevance in heterogeneous catalysis.

6.3.1 A potential energy diagram

Potential energy profiles for chemical adsorption on a metal substrate turn out to be most informative for the case of *dissociative* chemisorption, and so it is this type of adsorption that we shall concentrate on here.

To begin, let us consider the general case of the dissociative chemisorption of a molecule X_2 on a metal, M. Figure 9 shows schematically a possible potential energy profile for the overall process:

$$X_2(g) + 2*_M \longrightarrow \begin{array}{c} X \quad\; X \\ | \;+\; | \\ *_M \;\; *_M \end{array} \qquad (15)$$

where $*_M$ represents an adsorption site on the metal surface. The profile is often referred to as a **Lennard-Jones plot**, after J.E. Lennard-Jones, who first visualized dissociative chemisorption in this way. In Figure 9 the zero of potential energy is chosen to correspond to the situation in which the molecule X_2 is at an infinite distance from the metal surface. The first thing to notice about the plot is that it portrays *two* potential energy curves. The continuous line labelled $X_2(g) + M$ is just the potential energy curve for the *physical adsorption* of molecular X_2: it is very similar in character to that in Figure 8 – for the physical adsorption of molecular hydrogen on a metal. The other continuous line, labelled $2*_M + X(g) + X(g)$, represents the potential energy profile for *two atoms* of X as they approach the metal

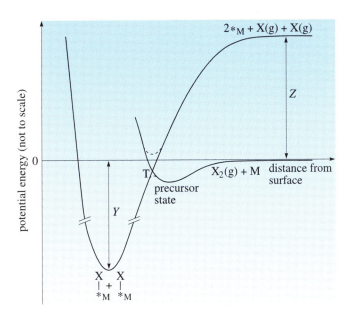

Figure 9 Schematic potential energy plot for the dissociative chemisorption of a molecule X_2 on a metal, M. The plot is not to scale and is drawn in an exaggerated fashion to highlight the main features.

surface and become chemically adsorbed. It is the formation of two $X-*_M$ chemical bonds that leads to the deep potential energy minimum at a distance much closer to the surface than in physical adsorption.

■ What does the energy difference labelled Z in Figure 9 represent? (For simplicity, ignore the effects of zero-point vibrational energy.)

■ It represents the bond dissociation energy of a *free* molecule of X_2, $D(X-X)$.

■ To what does the energy difference labelled Y correspond?

■ To a good approximation, it represents the enthalpy change for the process described by equation 15, that is the enthalpy change for chemical adsorption, ΔH_{ca}. (Note very carefully that equation 15 describes the interaction between X_2 *molecules* in the gas phase and the metal surface to give *atoms* of X on the surface.)

■ According to Figure 9, is chemical adsorption an exothermic or endothermic process?

■ From Figure 9, $H(2X-*_M) < H(X_2(g) + 2*_M)$, so $\Delta H_{ca} < 0$: the process is exothermic, as you might expect. (We return to this point in Section 6.3.2.)

A key feature of the Lennard-Jones plot is that the two potential energy curves cross one another: in Figure 9 the region where they cross is labelled T. For reasons that we must ask you to accept, this means that the two curves will 'mix' with one another and become rounded in the crossing region as indicated by the dashed curves.

Can you see the significance of this mixing?

Suppose a molecule of X_2 approaches the metal surface and becomes physically adsorbed: this physically adsorbed state is referred to as a **precursor state** for chemical adsorption. Providing the molecule can then obtain sufficient energy to take it close enough to the surface to reach the crossing region, T, it can then transfer to the $2*_M + X(g) + X(g)$ curve and *so become chemically adsorbed in the dissociated state*. As shown in Figure 9, there is a small energy barrier between the precursor state and the chemically adsorbed state. However, it must be remembered that the diagram is schematic; for instance, the physically adsorbed molecule might pass straight over to the chemisorbed state *before losing* its energy of physical adsorption, in which case the experimentally observed activation energy for chemical adsorption would be expected to be close to zero. Indeed, such behaviour is often observed for adsorption systems: typically those involving stringently cleaned metal surfaces and gases such as oxygen, hydrogen and nitrogen. It is thus a common convention to *depict* systems involving **non-activated chemisorption** by a Lennard-Jones plot in which the intersection of the two potential energy curves occurs *below* the energy of the free molecule (arbitrarily chosen as zero in our case). Figure 9 thus represents a non-activated chemisorption process.

To return to a specific example, let us now consider the dissociative chemisorption of hydrogen on a metal, such as nickel:

$$H_2(g) + 2*_{Ni} \longrightarrow \begin{matrix} H & & H \\ | & + & | \\ *_{Ni} & & *_{Ni} \end{matrix} \qquad (16)$$

■ The Lennard-Jones plot for this process is shown in Figure 10. Is this a non-activated chemisorption process?

▪ No. The potential energy curves cross at an energy *above* that of the free hydrogen molecule.

It should be clear from Figure 10 that there is now a relatively significant energy barrier between the physically adsorbed precursor state and the final chemisorbed state, with the *transition state* for chemical adsorption lying in the region of intersection, T, of the two potential energy curves. Chemical adsorption is now an **activated process**, and it follows that one reasonable way to depict the **activation energy for chemical adsorption** is to represent it by the height of the potential energy barrier above *the potential energy of the free molecule* (arbitrarily assigned zero in our case). This barrier is labelled E in Figure 10,* and it should be clear that in general its value will be sensitive to the precise shapes and relative positions of the two curves in a Lennard-Jones plot.

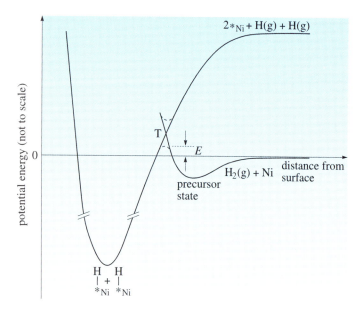

Figure 10 Schematic potential energy plot for the dissociative chemisorption of hydrogen on nickel metal. Again the plot is not to scale and is drawn in an exaggerated fashion to highlight the main features.

*In some cases the activation energy for a dissociative chemisorption process is measured on a Lennard-Jones plot from the bottom of the potential energy well for the physically adsorbed precursor state. However, the more usual convention is to adopt the description as given above (there is also another advantage to this convention as you will discover in SAQ 9). But it should always be kept in mind that the Lennard-Jones plot is no more than a *pictorial representation* of a rather complicated process.

For the adsorption of hydrogen on nickel it turns out that the activation energy for chemical adsorption is quite small: certainly less than $10\,kJ\,mol^{-1}$. But notice that if it had been necessary for the hydrogen in the gas phase *to dissociate before it could be chemically adsorbed*, then this would have required a significant expenditure of energy: equivalent to the bond dissociation energy of H_2, that is $432\,kJ\,mol^{-1}$. (Notice also that these numbers indicate emphatically that Figure 10 is not to scale!)

The attraction of the Lennard-Jones picture of dissociative chemisorption is that it provides a simple way of both visualizing, and explaining in energetic terms, how a molecule may dissociate on a surface without prior dissociation in the gas phase. Intrinsic to the model is a physically adsorbed precursor state. Admittedly this picture is simplified; for instance, it does not take into account the fact that the energy barrier to adsorption probably includes a contribution due to 'stretching' a molecule to match the distance between adsorption sites. Nevertheless, the idea of physical adsorption, followed by chemical adsorption, is a powerful one – and certainly useful in discussions of catalysis. This transition can be visualized in a schematic way as shown in Figure 11.

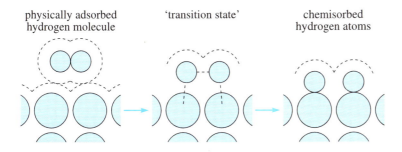

Figure 11 A simplified pictorial representation of the chemisorption of a hydrogen molecule on a metal surface.

STUDY COMMENT   You should try to work through the following SAQs before reading on. SAQ 10 in particular illustrates how a Lennard-Jones plot can provide valuable insight into the catalytic activity of different surfaces.

SAQ 9   Under certain conditions the enthalpy change for chemical adsorption, ΔH_{ca}, for hydrogen on nickel is $-96\,kJ\,mol^{-1}$. Use Figure 10, in conjunction with the values given in the text above, to determine (a) the activation energy for the *desorption* of hydrogen *molecules* from a nickel surface, and (b) the dissociation energy of the chemical bond between a nickel surface site and an adsorbed hydrogen atom, that is $D(H-*_{Ni})$.

SAQ 10   Considerable research work has concentrated on the synthesis of ammonia from its elements using an iron catalyst. One study (G. Ertl, M. Weiss and S.B. Lee, *Chemical Physics Letters*, January 1979) reported the dissociative chemisorption of nitrogen gas on a particular face of a single crystal of iron, prepared *both* in a clean state *and* in a state in which it was doped with potassium. The experiment (which was of considerable complexity) provided a great deal of quantitative information, one aspect of which can be illustrated by a Lennard-Jones type potential energy plot (Figure 12, overleaf). What role does potassium play in the dissociative chemisorption of nitrogen on iron?

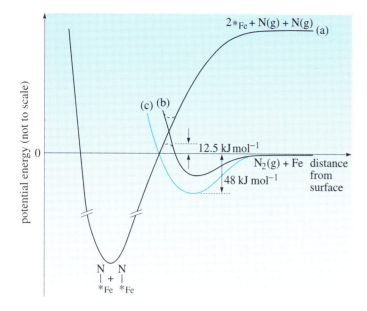

Figure 12 Schematic potential energy plots for the dissociative chemisorption of nitrogen gas on a particular crystal face of iron. Curves (a) and (b) are for a clean surface and curves (a) and (c) are for the same surface but doped with potassium. (For clarity, the 'rounding' of curves (a) and (c) where they intersect has been omitted.)

6.3.2 The enthalpy change for adsorption

STUDY COMMENT This Section includes a description of methods for measuring heats of adsorption, essentially to help illustrate how the enthalpy change for adsorption varies with the amount of adsorbate already present on a surface.

We have already indicated that a knowledge of the magnitude of the enthalpy change for an adsorption process provides an important method of distinguishing between physical and chemical adsorption: values greater than 40 kJ mol^{-1} are generally accepted to be indicative of the latter.* The enthalpy change for adsorption is, however, also of fundamental interest in chemical adsorption studies. It provides a measure, albeit indirectly, of adsorbate–substrate bond strengths, and even though it is a macroscopic quantity, it provides a good deal of information about surface heterogeneity. Here, our main concern will be with aspects that are of direct relevance in catalysis.

In simple terms – although we shall modify this view shortly – the enthalpy change for chemical adsorption refers to the enthalpy change at a fixed temperature in a reaction that can be represented schematically as:

$$\text{adsorbate(g)} + n* \longrightarrow \text{adsorbed chemical species} \qquad (17)$$

where the adsorption process can be either associative or dissociative.

SAQ 11 In most circumstances it is found that the entropy change for chemical adsorption, ΔS_{ca}, has a negative value. Bearing this information in mind, see if you can suggest a simple thermodynamic argument to rationalize the statement that 'chemical adsorption is virtually always an exothermic process'.

It may seem from what has been said so far that the enthalpy change observed when a small amount of substance is chemically adsorbed (at a fixed temperature) on a relatively large amount of substrate, should be the same irrespective of the state of the surface; that is, whether the surface is initially clean or already covered to some extent by adsorbed species. In fact this is rarely the case, and enthalpy changes for chemical adsorption (or physical adsorption for that matter) are found to depend

---

* It must be said, particularly in the light of results obtained by modern techniques – such as those described in Block 6 – that this simple, but practical, classification is not without uncertainty. The 'transition' between chemical and physical adsorption is best regarded as a continuous one, with the formation of a chemical compound at one extreme. Also, adsorbate species are often mobile on a solid surface and so, apart from adsorption strength, the nature of this mobility can often be crucial to the functioning of a catalyst.

markedly on surface coverage. This is a very important observation. Before we examine it in more detail, however, a slight digression to define surface coverage in a quantitative way will prove useful, both now and later.

It is usual to express surface coverage as a **fractional surface coverage**, θ:

$$\theta = \frac{\text{number of adsorption sites occupied}}{\text{number of adsorption sites available on a clean surface}} \tag{18}$$

Experimentally, this quantity is determined as the ratio of the amount (expressed in terms of the number of *molecules*) adsorbed under given conditions of temperature and pressure to the *monolayer capacity* of the surface – the amount contained in a single complete layer, either physically or chemically adsorbed. In the particular case of adsorption on single crystal surfaces when the exact surface structure is known (see Block 6), a more precise definition of θ becomes possible. It can be taken as the ratio of the number of molecules adsorbed to the number of substrate atoms in the top layer of the solid (both quantities relating to unit surface area). The amount of substance adsorbed can be determined using a variety of techniques: for example, direct weighing (modern balances can detect mass changes down to 10^{-11} kg), the measurement of pressure changes, or indirect methods that depend on the measurement of some physical property (the electrical conductivity for instance) of the substrate that is changed in a way proportional to the amount adsorbed.

Clearly, if an enthalpy change for adsorption can depend on the fractional surface coverage, we need a more refined method of dealing with this quantity. The problem is best highlighted by briefly considering how to measure enthalpy changes for adsorption in practice.

One direct method is to use *calorimetry*: enthalpy changes for chemical adsorption can be measured in an analogous way to those for ordinary chemical reactions. The form of the metal substrate, for instance polycrystalline powder, wire or evaporated film, largely determines the type of calorimeter to be used. But all calorimeters have in common an overall requirement to measure accurately the very small temperature changes (often no more than small fractions of a degree) that result from the adsorption of only small amounts of gas. A schematic view of a fairly simple type of calorimeter is shown in Figure 13: it is designed to be used under *adiabatic* conditions, that is the heat flow to or from the adsorption system is kept as close to zero as possible.

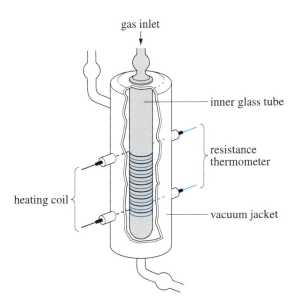

Figure 13 A schematic view of a calorimeter designed for the measurement of enthalpy changes for the adsorption of gases on evaporated metal films. The very thin-walled inner glass tube is covered on its inner surface with an evaporated metal film. The resistance thermometer (platinum or tungsten wire) is such that resistance differences corresponding to temperature changes as small as 10^{-4} K can be detected. The heating coil is for calibration purposes. Cold water can be passed through the outer vacuum jacket during the deposition of a metal film, and the jacket can be evacuated to reduce heat losses during a measurement.

At this stage we also need to be clear about a convention commonly used in adsorption studies. For an adsorption process, which we will now assume to be invariably exothermic, it is usual to quote the *magnitude* of the enthalpy change as a **heat of adsorption**; for example, if the enthalpy of adsorption is -80 kJ mol^{-1}, then the heat of adsorption is quoted as 80 kJ mol^{-1}. To be consistent with the literature we shall now adopt this practice.

Suppose in a typical experiment that a small amount of a gas, A, is admitted to a metal substrate and n_A moles of it are adsorbed at constant volume and temperature giving some fractional coverage θ. The measured heat of adsorption (or enthalpy change) will represent a total value over the fraction of the surface covered. For this reason it is called an **integral heat of adsorption** and given the symbol Q_i: typically, it is measured in kilojoules. Suppose now an additional very small quantity of gas, Δn_A, is admitted, increasing the amount adsorbed to $(n_A + \Delta n_A)$, and liberating heat ΔQ_i. *Providing* that the magnitude of Δn_A is small compared with that of n_A, then a **differential heat of adsorption**, q_d, can be calculated from the ratio:

$$q_d = \Delta Q_i / \Delta n_A \tag{19}$$

Typically, it is measured in kJ mol^{-1}.

An alternative approach is to measure the integral heat of adsorption as a function of the amount adsorbed. In this case, the differential heat of adsorption for a *particular amount adsorbed* is given by the slope of a tangent (that is the differential) at this point on a plot of Q_i versus n_A; that is

$$q_d = dQ_i/dn_A \tag{20}$$

Can you see the significance of the differential heat of adsorption?

As already hinted, it represents the heat of adsorption (expressed in molar terms) that would be observed when the surface is already covered with a given amount of adsorbate; or in other words, it is the heat of adsorption *at a particular fractional surface coverage.**

We are now in a position to look at a few experimental results. Figure 14 shows how experimental differential heats of adsorption vary as a function of fractional surface coverage for the adsorption of hydrogen on a selection of evaporated metal films. It is apparent that in all cases the differential heat of adsorption varies, often markedly, with fractional surface coverage and tends to *decrease* with increasing coverage of the surface. Indeed, such behaviour is found for many other chemical adsorption systems employing polycrystalline substrates, and also in other studies using single crystals as substrates. In the latter case indirect methods, rather than calorimetry, are used to determine the differential heat of adsorption as a function of coverage: an example is given in Figure 15.

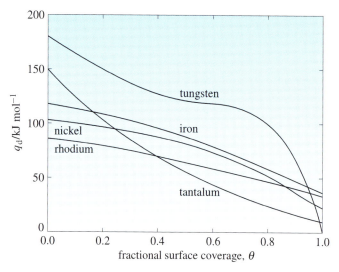

Figure 14 The variation of calorimetrically determined differential heats of adsorption as a function of fractional surface coverage for the adsorption of hydrogen on a selection of evaporated metal films at room temperature. (For clarity, smooth curves have been drawn through the experimental points for each metal.)

*Strictly, this is not quite true. The heat of adsorption at a given fractional surface coverage is a fundamental thermodynamic quantity called the *isosteric heat of adsorption*, q_{st}. Rather complex thermodynamic argument reveals that $q_{st} = q_d + RT$. The difference between the two heats is small (at 300 K, say it is of the order of 2.5 kJ mol^{-1}) and usually, from an experimental viewpoint, does not warrant distinction except perhaps in the case of physical adsorption studies.

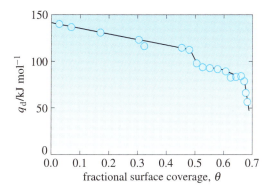

Figure 15 The variation of the differential heat of adsorption for carbon monoxide, CO, on the Pt(111) face of a single crystal of platinum as a function of fractional surface coverage. (The notation for single crystal faces is discussed in Block 6.)

Can you suggest any reasons for the observed behaviour?

In general there are several possibilities. First, **surface heterogeneity**, the causes of which were briefly outlined towards the end of Section 3, will give rise to a distribution of binding sites of different energy, particularly for polycrystalline substrates. The most active sites, which give rise to high differential heats of adsorption, will be covered initially, and then, with increasing coverage, the less active sites will be occupied with a consequent fall in the differential heat of adsorption. Second, with increasing surface coverage the likelihood of adsorbate–adsorbate interactions will increase; this is often referred to as a '(self-)induced heterogeneity'. Such interactions are generally repulsive in nature and so they effectively *decrease* the activity of adjacent unoccupied sites, and hence contribute to a lowering of the differential heat of adsorption. Finally, it is also possible that the nature of the bonding between the adsorbate and the surface may change at a particular surface coverage.

The information that the measurement of differential heats of adsorption as a function of fractional surface coverage provides about surface heterogeneity is important in its own right in chemical adsorption studies. But, as we shall see later, it also has a significant role to play in the formulation of kinetic models of surface-catalysed reactions.

SAQ 12 Studies of the rate of the dissociative chemical adsorption of nitrogen gas on an iron surface indicate that the activation energy for the process increases continuously with increasing fractional surface coverage. Can you suggest why this should be the case? [*Hint* Consider how the Lennard-Jones-type potential energy plots for the dissociative chemical adsorption of nitrogen will vary with surface coverage.]

6.3.3 Thermal desorption spectroscopy

Over the past few decades **thermal desorption spectroscopy** has established itself as an important tool in chemical adsorption studies. One reason for this is that it provides a means of measuring desorption rates and hence the activation energy for desorption; the latter quantity of course sets an upper limit to the heat of adsorption (cf. Figure 10 and the answer to SAQ 9). Another reason, and the one we wish to outline briefly here, is that it provides a direct method of distinguishing the different types of chemical adsorption site available on a surface.

The general procedure is to allow a gas at low pressure (often less than 0.01 Pa) to flow past a clean filament or crystal surface (usually of a metal), held at a fixed temperature, for a given time so that a certain amount is adsorbed. The sample is then heated in a controlled manner over a wide temperature range while the vessel in which it is contained is continuously pumped. The basic observation is that there is a pressure surge when a temperature *region* is reached at which the activation energy for desorption is 'overcome' and pressure changes due to desorption become observable.

An example of a **thermal desorption spectrum** is given in Figure 16. The appearance of each individual pressure peak can be explained (qualitatively) by the following argument. As the temperature is raised the rate of desorption will increase, *but* at the same time the amount of material available to be desorbed *must decrease*. If the reaction vessel had been *closed*, desorption into the fixed volume would have produced a 'step' in the desorption spectrum. With continuous pumping, however, the desorbed gas is eventually removed again and instead a pressure peak appears with a maximum at some characteristic temperature.

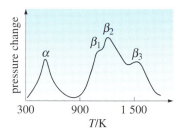

Figure 16 The thermal desorption spectrum of carbon monoxide, CO, on the surface of a particular crystal face of tungsten.

Qualitatively, what do you think is the significance of the several pressure peaks in Figure 16?

Clearly, there must be several different types of binding site on the tungsten surface with different desorption energies. Peaks occurring at the lowest temperatures will correspond to sites with the smallest activation energies for desorption (and hence smallest heats of adsorption). In fact, more detailed analysis reveals that the weakly bound state – labelled α in Figure 16 – probably corresponds to a linear W—CO bonding, whereas the strongly bound β states (with desorption activation energies ranging up to 360 kJ mol^{-1}) probably correspond to 'partial' dissociation of CO at various sites. The interpretation is not without ambiguity, but it does indicate the complexity of the chemical adsorption of a reasonably simple molecule on just one particular crystallographic plane of a given metal.

The conditions under which most thermal desorption studies are carried out – high vacuum and well-characterized substrates – are far removed from those used in heterogeneously catalysed reactions. Nonetheless, the investigations, particularly those carried out in recent years with single-crystal surfaces, do invariably indicate the involved nature of the process of chemical adsorption – especially in the case of those substrates with catalytic activity.

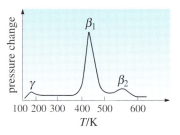

Figure 17 The thermal desorption spectrum of hydrogen on a particular face of a single crystal of tungsten.

SAQ 13 Figure 17 shows a thermal desorption spectrum of hydrogen on a particular face of a single crystal of tungsten. Noting the temperature range over which the spectrum was recorded, can you qualitatively suggest what types of surface binding the peaks probably correspond to?

STUDY COMMENT The following SAQ provides you with an opportunity to summarize for yourself some of the key differences between physical and chemical adsorption. Try to do this – and check your answer with ours – before going on to Section 6.4.

SAQ 14 Various criteria can be used to distinguish between physical and chemical adsorption. A convenient, and concise, way of presenting these criteria is in the form of a table with headings as indicated below:

Criterion	Chemical adsorption	Physical adsorption
(i) forces involved		
(ii)		
(iii)		
(iv)		

See if you can fill in the table by writing down criteria (in no more than a few words) and then indicating how each applies to both chemical adsorption and physical adsorption.

6.4 Summary of Section 6

1   Adsorption is the process by which atoms or molecules are attached to a solid surface. The material that adsorbs is the adsorbate and the underlying solid is the substrate (adsorbent).

2   Adsorption occurs at solid surfaces because the surface atoms are not fully coordinated and consequently exert a net attractive force on atoms or molecules with which they are in contact. The nature of adsorption on metallic and ionic surfaces makes solids of these types particularly suitable for the catalysis of industrially important reactions.

3   Generally, adsorption phenomena are of two types: physical adsorption and chemical adsorption. (Table 10 in the answer to SAQ 14 summarizes the main differences.)

4   Physical adsorption involves weak intermolecular forces, the enthalpy change for physical adsorption is typically smaller than $-40\,kJ\,mol^{-1}$, and the chemical character of an adsorbate molecule is hardly changed.

5   Chemical adsorption – dissociative or associative – involves the formation of chemical bonds. Enthalpy changes for chemisorption typically lie in the range from -40 to $-400\,kJ\,mol^{-1}$, and the chemical character of an adsorbate molecule is substantially altered.

6   The Lennard-Jones potential energy representation of chemical adsorption depicts in simple terms how a molecule, such as hydrogen, can be dissociatively adsorbed, without prior dissociation in the gas phase, via a physically adsorbed precursor state of the molecule.

7   In virtually all systems chemical adsorption is an exothermic process.

8   The differential heat of adsorption, which can be determined by calorimetric and other methods, is usually found to decrease with increasing fractional surface coverage; as such it provides a means of characterizing surface heterogeneity.

9   Thermal desorption spectroscopy provides information on the type and stability of species chemically adsorbed on a surface, and on the types of site available for adsorption.

7 CHEMICAL ADSORPTION AND CATALYSIS

The belief that chemisorption of one or more of the reactants is involved, as an intermediate step, in essentially all solid-catalyzed reactions leads to the hope that an understanding of chemisorption phenomena on catalysts would illuminate and clarify the mechanisms of catalytic action. A vast literature exists on chemisorption and the relationships between chemisorption and catalysis, but most generalizations as have emerged must be hedged with such qualifications that little can be said in summary that is helpful without being misleading.

(Charles N. Satterfield, *Heterogeneous catalysis in practice*, McGraw-Hill, 1980).

The above quotation summarizes concisely the difficulty of describing, without going into detail for each system, the relationship between chemical adsorption studies and catalysis. Nonetheless, as you will see in due course, it is possible to make a few useful generalizations, particularly against the background of the material developed in Section 6.

It is also pertinent to note at this stage one other extremely important aspect of chemical adsorption, which we have hardly mentioned so far – the use of modern spectroscopic and crystallographic techniques to determine the nature of chemically adsorbed species and the processes that they undergo at the atomic level. At first sight, it would seem that these methods should be of key significance in establishing the mechanism of a catalysed reaction. However, the conditions that most 'surface science' techniques require – for example, atomically clean crystal surfaces under high vacuum conditions – are far removed from those used in 'real' catalysis. As a consequence, the information that they provide is often of debatable value. It is for this reason that we discuss them separately, and in some detail, in Block 6.

It is obvious from Table 1 that solid catalysts are a diverse set of materials; they range from simple metals to substances of complex composition. This strongly suggests, as we would expect, that it is the chemical nature of a solid, or more precisely the chemical nature of its surface, that determines its ability to act as a catalyst in a given reaction. Clearly, if we can classify solid catalysts in some way, this will help us to order our discussion.

7.1 A classification of solid catalysts

One convenient way to classify solid catalysts is in terms of their principal applications. This is done in Table 2, which shows that, depending on the type of chemical reaction, there is a particular class of solid catalysts that are active. Thus transition metals are good catalysts for hydrogenation reactions; as we shall see, this is because their surfaces interact in a special way with hydrogen molecules. Similarly, some metal oxides are excellent oxidation catalysts because they have a strong surface interaction with oxygen and other molecules. Other oxides, such as alumina, silica, and magnesia, have little affinity for oxygen, but their surfaces readily interact with water and so they are useful for dehydration reactions. Finally, a number of solids have acidic surfaces and can catalyse reactions that are very similar to those catalysed by mineral acids in homogeneous solution.

Table 2 Classification of heterogeneous catalysts.[a]

Catalyst class	Type of chemical reaction	Examples
metals	hydrogenation hydrogenolysis[b]	iron, nickel, palladium, platinum, copper
	oxidation	silver, platinum
metal oxides	dehydration	alumina, magnesia, silica[c]
	oxidation	vanadium(V) oxide, nickel oxide, zinc oxide, chromium oxide, complex metal molybdates, multimetallic oxide compositions
acids	isomerization alkylation cracking	silica–aluminas, zeolites in acid form

[a] The classification is by no means complete. There are exceptions, and reactions that it is not possible to classify, but this is in the nature of the subject.

[b] Hydrogenolysis (literally, 'splitting with hydrogen') is the addition of hydrogen across a single bond to cause splitting into two molecules, for example $CH_3-CH_3 + H_2 \longrightarrow 2CH_4$; it is also termed *hydrocracking*.

[c] Silica, SiO_2, is not a metal oxide, but it is convenient to include it in this table as such.

7.2 Metals

The ability of metals to catalyse a given chemical reaction varies quite markedly: some are active, some are not. One way to rationalize this behaviour is to consider the adsorptive properties of metals. It is found for most metals (gold being an exception) that the strength of chemical adsorption of simple gases and vapours of practical interest decreases in the following sequence: $O_2 > C_2H_2 > C_2H_4 > CO > H_2 > N_2$. Some metals have very active surfaces and can chemically adsorb all of these species; others are less active and can adsorb only oxygen. These observations lead to the idea of grouping metals according to their ability to chemically adsorb simple molecules. If all considerations of fine detail are waived, and the criterion for chemical adsorption taken to be that 'it occurs if it can be detected at a pressure of say 100 Pa and room temperature', then Table 3 can be constructed.

> If you look at the Periodic classification of the elements in the *Data Book* for the Second Level Inorganic Course, can you discern any general trend in the way the metals have been divided into groups in Table 3?

In most cases, the metals of a particular group occur in a certain area of the Periodic classification. Metals of groups A and B_1 are transition elements. Group B_2 also contains two transition elements, but these often show anomalous (or not measured, or not understood) behaviour, hence their separate grouping. The majority of metals in groups C, D and E come either before or after the transition series (Ag, Au, Zn and Cd come at the end of the transition element series). Overall the message is clear: high chemical adsorption ability is mostly confined to metals within the transition element series. Why?

The question posed above is not easy to answer. Certainly, atoms of transition elements are different from those of other elements in that they can have one or more unpaired *d* electrons in their outermost electron shells. Of course, for a metal we should consider the electronic properties of its atoms in bulk; however, according to *band theory*, which was discussed in the Second Level Inorganic Course, the electrons in a metal retain much of the character they possess in the isolated atoms and so it is still legitimate to talk about *d* electrons in the solid. (But whereas in an isolated atom each energy level is discrete and single-valued, in a metal crystal each energy state has a band of permitted values.) Thus one explanation goes along the lines that chemical adsorption on metals involves covalent bonding with unpaired electrons in *d* orbitals (or, more strictly, partly filled *d* bands). Although plausible, it turns out that this theory is in fact of limited value, and has been superseded by more sophisticated electronic theories. But as yet, none offers any simple explanation.

Table 3 Classification of metals according to their abilities in chemical adsorption: + indicates strong chemical adsorption, ± weak chemical adsorption, and − chemical adsorption not observed.

Group	Metals	O_2	C_2H_2	C_2H_4	CO	H_2	N_2
A	Ti, V, Cr, Fe, Zr, Nb, Mo, Ru, Hf, Ta, W, Os	+	+	+	+	+	+
B_1	Co, Ni, Rh, Pd, Ir, Pt	+	+	+	+	+	−
B_2	Mn, Cu	+	+	+	+	±	−
C	Al, Au[a]	+	+	+	+	−	−
D	Li, Na, K	+	+	−	−	−	−
E	Mg, Ag, Zn, Cd, In, Si, Ge, Sn, Pb, As, Sb, Bi	+	−	−	−	−	−

[a] Gold does not chemically adsorb oxygen.

Nevertheless, Table 3 does have a certain predictive value in catalysis. For instance, it indicates that catalysts for hydrogenation reactions, which require the adsorption of hydrogen, should be selected from groups A and B_1. But a word of caution. The table is constructed on the basis of simple qualitative observations; by contrast, catalytic reactions often have complex mechanisms that can depend critically on the presence of certain adsorbates in only very small amounts.

■ Which metals might be used to catalyse the synthesis of ammonia from its elements?

□ If we assume that a key step in the mechanism is the weakening of the strong bonding in molecular nitrogen by chemical adsorption, then only those metals in group A could be useful. This group includes iron.

The distinction between the adsorptive and catalytic properties of transition and non-transition metals is important. But the question still remains as to which is the 'best' transition metal to catalyse a given reaction. Perhaps not surprisingly, one answer is to rely on trial and error, but this is not to say that some useful guidelines cannot be discerned.

A good catalyst should be able to form substrate–adsorbate bonds of *intermediate strength*. By intermediate strength we mean that the bonds should be strong enough to ensure high surface coverage coupled with substantial rearrangement of the electrons in the reactant molecules, and yet not so strong as to cause a decrease in catalytic activity because adsorbed intermediates become immobilized on the surface. Immobile intermediates block the adsorption of new reactant molecules: in other words they *poison* the surface. The overall, but admittedly idealized, picture is that of catalytic activity increasing with strength of adsorption and then falling again: the situation is sometimes sketched schematically, as in Figure 18, in the form of a so-called **volcano curve**. (This picture really relates to the case in which there is a single reactant. When there are two reactants the situation is more complicated, but the same general principles apply.)

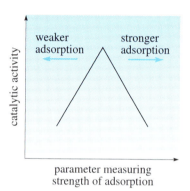

Figure 18 A sketch of a volcano curve.

But how do we measure the strength of adsorption? One suitable measure would seem to be the enthalpy change for adsorption, although we must remember, as discussed in Section 6.3.2, that this is quite a complex quantity. In particular, a single value cannot be quoted unequivocally for a given metal, because its magnitude will vary with surface coverage, or even with the type of crystal face exposed to the adsorbate. However, we must also keep in mind that our task here is rather empirical and so, as a compromise, we shall use values of differential heats of adsorption for polycrystalline metals determined at very low fractional surface coverages (essentially extrapolated to zero); these are referred to as **initial differential heats of adsorption**.

Figures 19 and 20 show how the initial differential heats of adsorption of hydrogen and oxygen vary on a selection of polycrystalline transition metals. You may notice that in some cases more than one value is plotted for a particular metal; this is not a reflection of the accuracy of the experimental measurements, but rather of the ill-defined nature of polycrystalline metal surfaces. This, in turn, means that results from samples prepared by different research groups can be expected to be close, but rarely identical.

For both hydrogen and oxygen, the magnitude of the initial differential heat of adsorption decreases in roughly a continuous manner across all three transition metal series, with the metals Co, Rh, Ir and Ni, Pd, Pt having the smallest values. How does this behaviour correlate with catalytic activity?

As an example let us take a reaction involving hydrogen, for which transition metals are widely used as catalysts. Figure 21 shows how the logarithm of the relative rate of hydrogenation of ethene,

$$C_2H_4(g) + H_2(g) = C_2H_6(g) \tag{21}$$

in the presence of various metals, varies with the initial differential heat of adsorption of hydrogen on the same metal. (Fewer adsorption data are available for ethene than for hydrogen, but what there are show the same trends.)

How can the information in Figure 21 be interpreted?

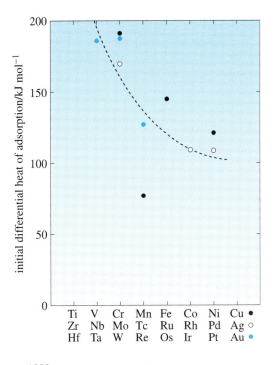

Figure 19 The initial differential heats of adsorption of hydrogen on various polycrystalline transition metals as substrates. (The curve has no significance other than helping to indicate the general trend.)

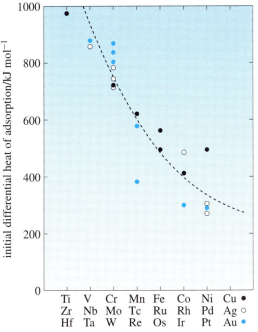

Figure 20 The initial differential heats of adsorption of oxygen on various polycrystalline transition metals as substrates. (Again the curve has no significance other than indicating the general trend.)

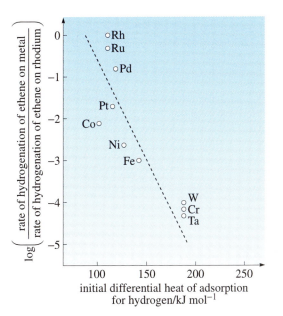

Figure 21 The variation of the logarithm of the relative rate of hydrogenation of ethene in the presence of various metals with the initial differential heat of adsorption of hydrogen on the same metal. The relative rate is expressed as the ratio of the rate of hydrogenation in the presence of a metal compared with the rate, under the same conditions, in the presence of rhodium.

It would seem, as expected, that the stronger the adsorption, as measured by the magnitude of the initial differential heat of adsorption, the slower the hydrogenation reaction. Indeed the effect is very marked: as Figure 21 shows, the rate of hydrogenation differs by as much as several orders of magnitude between one metal and another. It is still, however, contentious as to whether interpreting the data in terms of the right-hand part of a volcano curve is taking matters too far.

The correlation shown in Figure 21 is typical of the behaviour of transition metals in many other hydrogenation and hydrogenolysis (cf. Table 2) reactions. Invariably the metals Ru, Rh, Pd, Os, Ir and Pt are the most active, although their *relative* activities often vary quite markedly (and unpredictably) from one reaction to another. In very simple terms, the metals must evidently have 'intermediate' adsorption strengths for hydrogen. However, non-chemical factors also affect the choice of a catalyst; these metals are expensive and so economic considerations often dictate that less active, but cheaper, metals, such as nickel, are used in industrial hydrogenation processes.

If we turn, albeit briefly, to reactions involving oxygen, it turns out again that it is metals belonging to the 'platinum group' (Ru, Rh, Pd, Os, Ir and Pt), plus silver, that are the most efficient catalysts. Other metals tend to adsorb oxygen so strongly that they are converted into the oxide throughout their bulk and so are rendered inactive.

7.3 Metal oxides

If you look back to Table 2 you will see that metal oxides are divided into two groups depending on their function. In the first group are metal and semi-metal oxides of pre-transition elements (for example, alumina, magnesia and silica). These oxides are chemically very stable and have low electrical conductivity, so that they are good electrical insulators. They are essentially anhydrous, but readily sorb water, hence their role as catalysts in dehydration reactions.

In the second group are metal oxides of transition and post-transition metal elements. These oxides may contain a single metal (for example, vanadium(V) oxide) or be mixed oxides containing two or more metals (for example, bismuth molybdate-based catalysts). They all have ionic structures that are *non-stoichiometric* and, as you will see, the important property that oxygen may be transferred *to* or *from* their crystal lattices. Their electrical conductivities are low and they are *semi-conductors*: indeed, as a catalyst class they are often referred to as the **semiconducting oxides**. A description of the basic properties of semiconductors was given in the Second Level Inorganic Course.

Semiconducting oxides excel as oxidation catalysts. Most, but not all, of the industrial catalysts of this type are mixed oxides containing two (or more) metals: a few examples are given in Table 1. The beauty of these compounds is that under suitable conditions they can act as highly selective, *partial* oxidation catalysts; for instance, they allow cheap (a relative term) hydrocarbons to be converted, by incorporation of oxygen, into valuable organic chemicals. Pure oxides, that is those containing a single metal, generally catalyse total oxidation of hydrocarbons to carbon dioxide and water (sometimes referred to as 'deep' oxidation). For this reason they are not so interesting industrially, but they are important in the control of atmospheric pollution.

Correlations between the chemical and electronic properties of semiconducting oxides and their catalytic activity are difficult to make, partly because of their often complex composition. In this Section therefore we are forced to rely on a mix of general observations, coupled with specific examples, to give some flavour of this very interesting (and often industrially secretive) area.

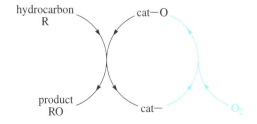

Figure 22 A redox mechanism for the oxidation of a hydrocarbon, R.

The behaviour of many oxidation catalysts can be described within the framework of a **redox mechanism**. In schematic outline the mechanism can be represented by a catalytic cycle (Figure 22), in which the two main steps, shown in black and blue, correspond to equations 22 and 23.

Step A (in black) The hydrocarbon (or related substance), which we shall simply call R, reacts with the catalyst surface at sites containing oxygen, represented schematically as cat—O, so that these sites are reduced (that is give up oxygen):

$$R + \text{cat}-O \longrightarrow RO + \text{cat}- \tag{22}$$

where RO is taken to represent partially oxidized product(s).

Step B (in blue) The reduced catalytic sites, cat—, are then replenished by oxygen from the gas phase in reactions that can be *summarized* as:

$$\tfrac{1}{2}O_2 + \text{cat}- \longrightarrow \text{cat}-O \tag{23}$$

This overall picture is often referred to as the **Mars and van Krevelen mechanism**, because it was first applied in 1954 by these workers. They used it in a more specific form to explain the kinetics of the oxidation of sulfur dioxide, SO_2, over vanadium oxide catalysts, and later in studies of the partial oxidation of various hydrocarbons.

The mechanism as described makes no assumption about the character of the oxygen at the surface sites; these sites are merely labelled cat—O. But, in practice, it is the nature of this oxygen that is the key to the mechanism. In many cases it is fairly certain that it is in the form of *surface* oxide anions, O_s^{2-}, derived from the lattice of the bulk oxide. Indeed, one of the main concepts underlying the development of selective, partial oxidation catalysts (pioneered by James Idol and co-workers at the Standard Oil Company of Ohio starting around 1952) was that 'lattice oxygen' of a reducible metal oxide could serve as a more versatile oxidizing agent for hydrocarbons than molecular oxygen from the gas phase. The role of gas-phase molecular oxygen is to *replenish* catalyst oxygen vacancies. But it must be said that the lattice oxygen mechanism is not exclusive; other *chemically adsorbed* oxygen species such as O_2^- and O^- or oxygen atoms doubly bonded to metal atoms can, and do, play important roles in catalytic oxidation mechanisms.

To return to a specific example, the partial oxidation of propene, $CH_2=CHCH_3$, to give propenal (acrolein):

$$CH_2=CHCH_3(g) + O_2(g) \longrightarrow CH_2=CHCHO(g) + H_2O(g) + \text{other products} \tag{7}$$

in the presence of bismuth molybdate-based catalysts has been extensively studied, in considerable part because of the industrial importance of this type of catalyst. The overall features, but not specific details, of the mechanism are reasonably well understood.

■ The first step is the chemical adsorption of the hydrocarbon, propene. Can you recall from the discussion in Section 6.2.2 how this occurs?

■ The propene is dissociatively adsorbed to form a symmetric allyl radical and a hydrogen atom:

$$H_2C=CHCH_3(g) + 2* \longrightarrow H_2C\underset{*}{\cdots}\overset{H}{\underset{|}{C}}\cdots CH_2 + \underset{*}{\overset{H}{|}} \tag{12}$$

The next step is a surface reaction between the adsorbed allyl radical and a surface oxide ion, O_s^{2-}, in which a further hydrogen atom is abstracted and propenal is formed; it can be represented as:

$$H_2C\underset{*}{\cdots}\overset{H}{\underset{|}{C}}\cdots CH_2 + O_s^{2-} \xrightarrow{-H} H_2C=\overset{H}{\underset{|}{C}}-CHO + 2e \tag{24}$$

— partial oxidation

The electrons released lower the oxidation states of the metal cations in the catalyst. The final step is reoxidation, whereby gaseous oxygen becomes incorporated into the lattice as ions to replace those lost in reaction 24:

$$\tfrac{1}{2}O_2(g) + 2e \longrightarrow O_s^{2-} \quad (25)$$

This last process is undoubtedly complex and probably involves two closely related processes: (a) adsorption and activation of molecular oxygen to form species such as O_2^- (adsorbed) and O^- (adsorbed); (b) the incorporation of these species into lattice vacancies as O^{2-} ions. However, as indicated in equation 25, no matter what the mechanism, the overall process requires electron transfer *from* the solid. Useful oxidation catalysts therefore often have, incorporated into the bismuth molybdate structure, other metal cations, such as iron, that can 'easily' change their oxidation states in order to facilitate this process.

As the mechanism emphasizes, the catalyst must itself be easily reduced (to facilitate the oxidation of propene) and also be capable of subsequent reoxidation. The adsorption of oxygen from the gas phase and subsequent reoxidation of the catalyst are believed to take place at sites distinct from those active for the partial oxidation of propene. It is proposed that oxygen is transported in the form of oxygen ions through the catalyst bulk, from the site where it is adsorbed, to reoxidize the catalytically active sites. A compensating transport of electrons completes the cycle. One possible mechanism discussed in the literature suggests that sites on molybdenum are responsible for the oxidation of propene and sites on bismuth for the reoxidation of the catalyst. The complex structure of bismuth molybdate catalysts, however, makes it difficult to establish unequivocal relationships between structure and activity. Three phases, α, β and γ, having mole ratios of MoO_3 to Bi_2O_3 of 3, 2 and 1, respectively, are often discussed, but even then it is most likely that there are variations of structure within each of these single 'phases'. The whole story of partial oxidation catalysts is fascinating, but certainly far from complete in the open literature.

It is worthwhile mentioning that for *simple, rather than mixed*, metal oxides correlations *can* be made between semiconductor type and catalytic activity. These correlations can be expressed in terms of relatively straightforward chemical ideas, coupled with the ideas of band theory and impurity semiconductors that were discussed in the Second Level Inorganic Course. A comparison between the chemical and catalytic properties of zinc oxide, ZnO, and nickel oxide, NiO, provides a useful illustrative example.

Ordinarily, zinc oxide is non-stoichiometric because it contains a small excess of zinc atoms. In simplified terms, this excess of zinc (in its zero oxidation state) is able to furnish electrons to an empty conduction band of the metal oxide, thus giving electrical conductivity: in fact, zinc oxide is an **n-type semiconductor** (n stands for negative because there is an excess of electrons).

■ If adsorption of oxygen on zinc oxide involves the process:

$$O_2(g) + 4e = 2O^{2-} \text{ (adsorbed)} \quad (26)$$

what will tend to limit this process?

■ It will be limited by the extent to which excess zinc is present, and therefore will not be extensive.

Nickel oxide is also non-stoichiometric because it is ordinarily 'oxygen rich', in the sense that occasional Ni^{2+} ions are missing: electroneutrality is preserved by some of the nickel being present in the +3 oxidation state as Ni^{3+}. Again in simplified terms, these Ni^{3+} ions can withdraw electrons from a full band, leaving holes that permit electrical conductivity. In fact, it is common to talk as though these holes, which are positive relative to the negatively charged electrons filling the band, carry the electric current. For this reason nickel oxide is a **p-type semiconductor** (p for positive).

■ Will the adsorption of oxygen, by a process such as that described in equation 26, be limited on nickel oxide?

■ No. In principle it should proceed to monolayer formation, because the electrons for the process described by equation 26 can be provided from the slightly depleted band (or by the formation of further Ni^{3+}).

In summary, therefore, it would appear that p-type metal oxides should generally be better oxidation catalysts than n-type oxides, their different catalytic activities stemming from their different abilities to adsorb oxygen. But, as always in this subject, a word of caution is required. The suggested correlation must be treated with care because it must be remembered that it is the *surface* of the oxide that is important, and its electrical conduction properties may depend on the nature of the adsorbed species.

SAQ 15 The decomposition of nitrous oxide (dinitrogen oxide),

$$2N_2O(g) = 2N_2(g) + O_2(g) \tag{27}$$

is catalysed by a number of metal oxides. The initial step in the mechanism of the reaction involves fission of the N—O bond:

$$N_2O(g) + e = N_2(g) + O^- \text{ (adsorbed)} \tag{28}$$

Which type, n or p, of metal oxide catalyst would you expect to be more efficient for this reaction?

SAQ 16 The walls of self-cleaning household ovens contain a catalyst. Can you suggest what type, and what properties it must possess?

7.4 Acids

The **silica–aluminas** are one of the most important groups of compounds with acidic surfaces. They are widely used as cracking catalysts: see for example Table 1. The structures that give rise to acidity and consequent catalytic activity can be rationalized as follows.

Silica itself has a three-dimensional structure, which may be thought of as being built up from tetrahedral $[SiO_4]^{4-}$ units in which each oxygen is shared between two tetrahedra. If we now imagine that aluminium atoms are substituted into this lattice in place of silicon atoms, then acidic surface sites can arise in one of two ways. If the aluminium atom with oxidation state +3 is surrounded by four oxygen atoms, it can be thought of as forming the unit $[AlO_4]^{5-}$, so that for each aluminium atom introduced there is one excess negative charge, which must be balanced to preserve electroneutrality. If this is achieved by attachment of a proton (produced by the dissociation of water) then the surface is acidic because it conforms to the classical (**Brønsted**) ideas of acidity. Alternatively, the aluminium atom may remain three co-ordinate and so be able to act as an electron-pair acceptor, that is as a **Lewis acid.** (Concepts of acidity are discussed in the Second Level Inorganic Course.) The two possibilities are illustrated in Figure 23.

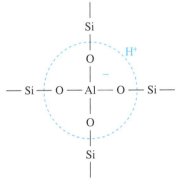

Brønsted acid

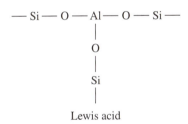

Lewis acid

Figure 23 Postulated structures of acidic surface sites for silica–alumina catalysts.

Despite the uncertainty as to which type of site is important under reaction conditions, it can be seen that the strongly acidic character of silica-aluminas arises from the 'mixing' of two compounds containing elements having different valencies: neither pure alumina nor pure silica shows much acidic character. Many other mixed compounds, for example $SiO_2.MgO$, $SiO_2.ZnO$ and $Al_2O_3.MgO$, have acidic surfaces for similar reasons.

Acid catalysts are very important industrially. The reactions they catalyse are complex and give rise to a wide range of products. In very broad outline it is thought that the formation of *carbocations* as intermediates in hydrocarbon reactions is a key feature in various catalytic mechanisms, for example:

$$—CH_2—CH_2—CH=CH_2 + \overset{H^+}{\underset{*}{|}} \longrightarrow —CH_2—CH_2—\overset{+}{C}H—CH_3 \tag{29}$$

carbocation

Carbocations are very reactive species and can undergo a variety of reactions, but the description of their chemistry is beyond the scope of this Course.

7.5 Zeolites

There are two major classes of aluminosilicates: the structurally *amorphous* silica–aluminas discussed above and their *crystalline* analogues, the **zeolites**. In recent years considerable attention has been devoted to the latter, very important class. Hence, although these are acid catalysts, we have given them a Section on their own.

> **STUDY COMMENT** There is a video sequence, *Zeolites* (video band 5), associated with this Section. It is perhaps best viewed after you have read about zeolite structure.

Naturally occurring zeolites have been known for about 200 years. The crystals come in many different shapes and forms, but – just like their amorphous cousins – at the atomic level the primary building blocks of all zeolites are SiO_4 and AlO_4 tetrahedra (Figure 24). These link together by sharing every oxygen to yield ordered, three-dimensional frameworks. As you will see, a special feature of zeolite structures is that they contain regular channels and cavities – known collectively as the **pore structure** – which are of molecular dimensions. This very fine pore structure, the nature and geometry of which effectively allows only those molecules of appropriate size and flexibility to penetrate into (or equally, to escape from) the interior of the crystal, gives zeolites an unusual degree of selectivity.

Figure 24 The base unit of all zeolites.

The pore systems of all zeolites contain rather loosely held cations (necessary to preserve electrical neutrality), which may be readily exchanged – hence their importance as water softeners (such as Permutit, a sodium-containing zeolite, which replaces the calcium in hard water by sodium, thereby rendering it 'soft') and purifiers. Both natural and synthetic zeolites also contain varying amounts of water: this is readily driven off by heating to yield anhydrous crystals suitable for catalysis. (Indeed, the name 'zeolite' – from the Greek for 'boiling stones' – was coined by A. F. Cronstedt in 1756, as a reference to the manner in which crystals of the mineral he discovered lost water visibly when heated.)

7.5.1 Zeolite structure

Zeolites vary widely in terms of their chemical make-up – the ratio of Al to Si atoms, for example, and the nature of the cations contained within them. Nevertheless, only a limited number of *framework structures* have been identified. Here we shall deal mainly with three of them: zeolite A, zeolites X and Y (which are isostructural) and ZSM-5 (**Z**eolite **S**ocony **M**obil – number **5**).

All zeolites contain *rings* of linked tetrahedral units, and the way these are built up into a three-dimensional framework is seen most clearly if the oxygen bridge between the tetrahedral centres (Al or Si atoms) is represented by a straight line. In this simplified scheme, the structure in Figure 25a, for example, becomes a hexagon (Figure 25b) – known as a *6-ring* – in which each intersection implies a tetrahedral atom. Typically zeolites contain rings with 4, 5, 6, 8, 10 and 12 tetrahedral units.

Typical of zeolite structures is the family of materials based on a secondary building unit composed of linked 4- and 6-rings forming a *truncated octahedron* or **sodalite unit** (Figure 26).

Primitive packing of these units yields the structure of sodalite itself (Figure 27a), a naturally occurring mineral. A similar stacking pattern, but with units separated from one another by oxygen bridges between the six 4-rings, generates **zeolite A** (Figure 27b). On the other hand, sodalite units linked via oxygen bridges through four of the eight 6-rings in a tetrahedral array, yields the framework structure of faujasite (Figure 27c). This is referred to as **zeolite X or Y**, depending on the ratio of aluminium to silicon.

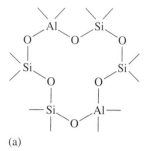

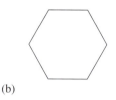

Figure 25 (a) A 6-ring containing two Al and four Si atoms; (b) shorthand version of the same 6-ring.

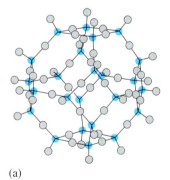

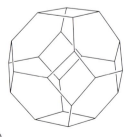

Figure 26 (a) A sodalite unit, comprising 24 tetrahedra arranged in six 4-rings and eight 6-rings; (b) representation of a sodalite unit as a truncated octahedron.

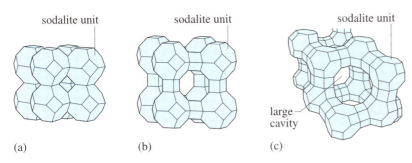

Figure 27 Zeolite frameworks built up from sodalite units: (a) sodalite; (b) zeolite A; (c) faujasite (zeolites X and Y).

You are not expected to memorize the structures shown here (or later), but simply to recognize how a three-dimensional pore system can be built up. Notice, in particular, that zeolite A and zeolites X and Y each contain large cavities (with diameters of 1 140 and 1 180 pm, respectively) linked together, the net effect being to generate considerable void space within the crystal (amounting to some 50% for these structures). It is within these cavities that sorption, diffusion and catalysis takes place; but access to them must be gained via the **windows** (or ports) that are apparent in Figure 27. Both the size and shape of these windows can thus influence access to the interior 'surface' of the crystal. For example, the sodalite window is a 4-ring that has a diameter (260 pm) too small to be of interest for catalysis.

ZSM-5 has a rather more complicated structure. It is not based on sodalite units, but instead can be generated by appropriate stacking of layers like the one shown in Figure 28a. In ZSM-5 this creates two types of channel system, which are interconnected (hence allowing diffusion in three dimensions), but not identical. In each case, the windows governing access are 10-rings (Figures 28a and 28b), but one system comprises nearly circular zigzag channels and the other elliptical straight channels (Figure 28c). The essential features of this structure are confirmed by high-resolution electron microscopy (Figure 29), which offers direct, 'real-space' proof of the regularity of the pores through which molecules must pass.

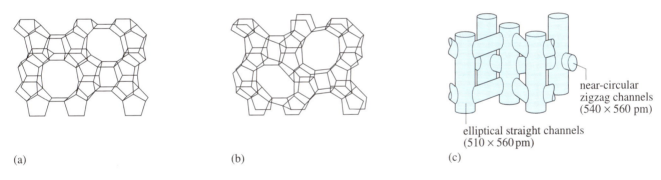

Figure 28 Structure of ZSM-5: (a) near-circular 10-ring entrances to zigzag channels; (b) elliptical 10-ring entrances to straight channels; (c) intersecting channel system.

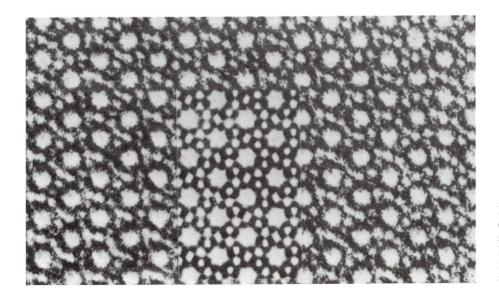

Figure 29 High-resolution electron micrograph showing the regular array of pores in ZSM-5. Inset is a computer-simulated image calculated on the basis of the structure in Figure 28a.

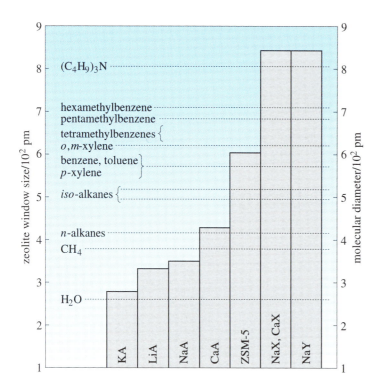

Figure 30 Comparison of port or channel dimensions of selected zeolites with effective molecular diameters.

Window sizes of these, and other, zeolite structures are correlated with the diameters of some typical sorbate molecules in Figure 30: the basis for using zeolites as molecular sieves is clear!

A word of caution, however: in practice, correlations like this are not always reliable. The window dimensions in Figure 30 were determined largely by X-ray crystallography, but molecules apparently larger than the reported port sizes are sometimes freely absorbed. Several factors may be involved, such as slight distortions to the crystal lattice, or uncertainties in the effective diameters of the more complex sorbate molecules.

Despite these reservations, Figure 30 does reveal that fine adjustments to the window size of certain zeolites can be achieved by varying the cation associated with the framework: compare, for example, the sodium and calcium forms of zeolite A, labelled NaA and CaA in Figure 30. As we said earlier, cations are present in zeolites to preserve electrical neutrality. In terms of their stoichiometry, zeolites can be thought of as being derived from silica, SiO_2, in which some silicon atoms have been replaced by aluminium: each such substitution contributes a formal charge of -1, which must be balanced by a corresponding number of positive charges. Thus the *general structural formula of zeolites* can be written $M_{x/n}\{(AlO_2)_x(SiO_2)_y\}.wH_2O$, where n is the oxidation state of the cation M and x is the number of Al substitutions in the unit cell (the term in curly brackets).

■ The unit cell of the framework structure of anhydrous zeolite A is $\{(AlO_2)_{12}(SiO_2)_{12}\}$. What total cationic charge is required?

▨ With -1 contributed by each Al centre, the total cationic charge must be $+12$, for example Na_{12}... or Ca_6... .

One form can quite easily be exchanged for another, for example sodium cations can be exchanged quite simply for other cations. In the particular case of zeolite A, it is known that *monovalent* cations (such as Na^+) can occupy sites *within* the 8-ring windows, and hence restrict entry to the large cavity shown in Figure 27b. Exchange of Ca^{2+} ions for Na^+ apparently removes the latter from these sites, and hence *increases* the free diameter of the window, as shown in Figure 30. Similar 'fine-tuning' can be achieved with other zeolites.

7.5.2 Active sites in zeolites

Like amorphous silica–aluminas (Section 7.4), the inherent catalytic activity of zeolites can be attributed to acid sites. In general, these sites are located both within the pore structure and on the outside surface: calculations from adsorption measurements suggest that the ratio of internal to external sites is typically about 100 : 1. At the same time, the very porous structures described earlier give zeolites total surface areas in the range 200–800 m^2 g^{-1}, which is among the highest of all commercial catalysts. It can also be noted that the catalytic activity of zeolites – certainly in so far as hydrocarbon transformations that proceed via carbocation intermediates are concerned – depends not only on the number of acid sites, but also on their acid strength. As we said in Section 7.4, the acid sites may be either Brønsted or Lewis in character (Figure 23). In the former case the proton is thought to form a bridging hydroxyl group:

$$\begin{array}{c} \text{H}^+ \qquad\qquad\qquad\qquad \text{H}^+ \\ | \qquad\qquad\qquad\qquad\qquad | \\ \text{O} \quad \text{O} \quad \text{O} \quad \text{O} \quad \text{O} \quad \text{O} \quad \text{O} \\ \diagdown\diagup\ \diagdown\diagup\ \diagdown\diagup\ \diagdown\diagup\ \diagdown\diagup\ \diagdown\diagup \\ \text{Si} \quad \text{Al}^- \quad \text{Si} \quad \text{Si} \quad \text{Al}^- \quad \text{Si} \\ \diagup\diagdown\ \diagup\diagdown\ \diagup\diagdown\ \diagup\diagdown\ \diagup\diagdown\ \diagup\diagdown \\ \text{O} \quad \text{O} \quad \text{O} \quad \text{O} \quad \text{O} \quad \text{O} \quad \text{O} \quad \text{O} \quad \text{O} \quad \text{O} \quad \text{O} \end{array}$$

Various attempts have been made to establish quantitative correlations between the acidity of zeolites and their composition, particularly in terms of the ratio of silicon to aluminium in the framework structure. Such correlations must be regarded with caution, because they inevitably presuppose a measure of compositional homogeneity throughout the structure. Nevertheless, on an empirical level, certain trends are apparent. For example, infrared studies on a range of zeolites (in their hydrogen forms) suggest that the O—H stretching frequency of the bridging hydroxyl group associated with Brønsted sites usually falls in the range 3 600–3 660 cm^{-1}. Sample values for three faujasite-like zeolites with different Si : Al ratios in the framework are given in Figure 31.

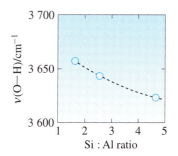

Figure 31 Variation of the O—H stretching frequency, v(O—H), with the framework composition of faujasite-like zeolites (the curve joining the points has no theoretical significance).

- What do these results suggest about the way the acid strength of Brønsted sites varies with composition?

- As the Si : Al ratio increases, the stretching frequency falls, implying a weakening of the O—H bond and an increase in acid strength.

Provided that the 'test' reaction is chosen with care, the rise in acid strength with increasing Si : Al ratio is paralleled by the relative catalytic activity of these zeolites. A similar pattern is shown by other families of zeolites: certainly forms rich in silica are generally found to contain very strong acid sites (but see the following SAQ).

SAQ 17 In a series of related zeolites, the catalytic activity (for a given test reaction) rises at first with increasing Si : Al ratio in the structure, reaches a maximum at a ratio of some 5–20 : 1, and then declines. Given the trends outlined above, can you suggest a possible explanation?

7.5.3 Shape-selective catalysis

As mentioned earlier, the acid sites in a zeolite are largely located within its internal structure, in a network of channels and cavities of molecular dimensions. The constraints this pore system imposes on the sorption and diffusion of reactant or product molecules, and on the local environment of the active sites themselves, are the basis for the often extraordinary selectivity of zeolite catalysts. This has been given the general name of **shape-selective catalysis**; it was first described in 1960, since which time three broad kinds of shape selectivity have been recognized.

Reactant selectivity is a direct consequence of the molecular 'sieving' properties of zeolites: it arises whenever one or more of the potential reactants is denied access to the active sites. A well-documented example is the dehydration of a mixed stream of the alcohols *n*-butanol (butan-1-ol, **1**) and *iso*-butanol (butan-2-ol, **2**), to form an alkene, over two different zeolites, one the wide-pore faujasite NaX (window

CH$_3$—CH$_2$—CH$_2$—CH$_2$—OH

1

CH$_3$—CH$_2$—CH—CH$_3$
                |
                OH

2

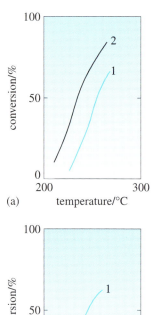

size about 850 pm) and the other zeolite CaA, which has much narrower pores (window size about 425 pm). Sample results are given in Figure 32: note that they are presented as percentage conversion after a given residence time, and hence reflect the relative rates of conversion of the two alcohols, as a function of temperature. Evidently, these are roughly comparable over NaX (Figure 32a), but there is negligible dehydration of *iso*-butanol over CaA (Figure 32b). In this case, the explanation is straightforward: both alcohols have ready access to the interstices of NaX, but the more bulky branched isomer (**2**) is effectively excluded from CaA; the minimal conversion that occurs presumably takes place on external sites.

A second kind of selectivity – **product selectivity** – arises when, among all the product species formed within the pore system, only those with the appropriate dimensions can diffuse out and appear as observed products. A molecule diffusing through a zeolite crystal is constantly under the influence of the zeolite surface, and this gives rise to a diffusion 'regime' – known variously as *restricted* or *configurational diffusion* – which is different in kind from that in a bulk fluid. As a result, rates of diffusion in zeolites can be many orders of magnitude smaller than those in the bulk fluid phase (a factor of 10^{-10} or so is not uncommon), and highly sensitive to slight variations in the size and shape of the diffusing molecule.

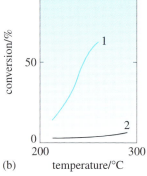

Figure 32 Percentage dehydration of butanols **1** and **2** over (a) zeolite NaX, and (b) zeolite CaA (both at 1 atm pressure and for a residence or contact time of 6 s).

The effect can be dramatic indeed – as exemplified by the alkylation of toluene (**3**) with methanol over the Mobil zeolite ZSM-5. The product is a mixture of xylenes (**4**, **5** and **6**), but it contains a disproportionately high percentage of just one isomer, *para*-xylene (**4**).

3

4, *para*-xylene

5, *meta*-xylene

6, *ortho*-xylene

The explanation lies with the slight differences in the size and shape of the isomers: *p*-xylene fits easily within the channels of ZSM-5, but the other isomers do not. (This is confirmed by the comparison of molecular dimensions in Figure 30). In consequence, the rates of diffusion of *meta*- and *ortho*-xylene are found to be some 10 000 times smaller than that of the *para*-isomer. Even if the primary product of alkylation is rich in the *m*- and *o*-forms, their retention within the structure allows further isomerization to *p*-xylene, which can exit rapidly from the crystal, thereby enhancing the observed yield of this isomer. This property of ZSM-5 has also seen commercial exploitation, allowing selective conversion of mixed xylene streams into the *para*-isomer, a vital precursor in the manufacture of Terylene.

A more subtle type of shape selectivity derives from the fact that reaction takes place within the channels and cavities of the zeolite crystals. If the spatial configuration of these intracrystalline voids is too restricted to allow a bulky transition state to form, then this introduces a further constraint on the reaction pathways available, in principle, to the reactants (or intermediate species derived from them).

An example of this so-called **restricted transition-state selectivity** is provided by a study of the relative rates of disproportionation (equation 30) and isomerization (equation 31) of *o*-xylene (**6**), over a variety of zeolites with different cavity sizes.

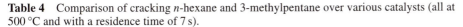

(30)

6   6   3   7

(31)

6   5   4

As Figure 33 shows, the rate of disproportionation relative to that of isomerization falls considerably as the cavity size is reduced, from zeolite HY (the hydrogen form of zeolite Y) to ZSM-5. This is presumably because disproportionation is a bimolecular process, requiring proper orientation of a relatively bulky transition state in a confined space, whereas isomerization simply requires consecutive *intra*molecular methyl shifts, each involving transition states not significantly larger than the resulting xylene.

One final point: shape selectivity evidently depends on the detailed microstructure of the zeolite, and on the strength, distribution and environment of the active sites within it. Thus, the factors outlined above are often not independent – and it is then necessary to invoke more than one type of shape selectivity in order to rationalize the observed distribution of products. For example, product selectivity also comes into play in the reaction of *o*-xylene over ZSM-5, enhancing the yield of *one* of the isomers, *p*-xylene, in the product stream, as expected from the discussion above.

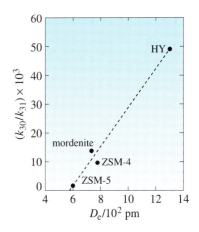

Figure 33 Variation in the relative rate constants for reactions 30 and 31, with zeolite cavity dimensions. (D_e is the effective diameter of the intracrystalline cavity.) ZSM-4 is another synthetic zeolite, whereas mordenite is a naturally occurring mineral.

SAQ 18 Table 4 contains sample results from experiments to compare the thermal cracking of two C_6 alkanes, 3-methylpentane (**8**) and *n*-hexane, over a variety of acid catalysts – two forms of zeolite A and a standard amorphous silica–alumina catalyst. The cracking reaction involves rupture of C—C bonds to produce a mixture of smaller hydrocarbons: the proportion of branched (*iso*-) to straight-chain C_4 and C_5 hydrocarbons in the products of *n*-hexane cracking are also recorded in Table 4. What do these results suggest about the shape-selectivity characteristics of zeolites NaA and CaA?

$$\text{CH}_3\text{CH}_2\text{CHCH}_2\text{CH}_3$$
$$|$$
$$\text{CH}_3$$

8

SAQ 19 When toluene is alkylated with methanol over ZSM-5, increasing the crystallite size from <0.5 μm to 3 μm roughly doubles the percentage of *p*-xylene in the product stream. Try to suggest a possible explanation.

Table 4 Comparison of cracking *n*-hexane and 3-methylpentane over various catalysts (all at 500 °C and with a residence time of 7 s).

Catalyst	Cracking conversion/%		Products from *n*-hexane	
	3-Methylpentane	*n*-Hexane	*iso*-C_4/*n*-C_4	*iso*-C_5/*n*-C_5
amorphous silica–alumina	28	12.2	1.4	10
zeolite NaA	<1	1.4	–	–
zeolite CaA	<1	9.2	<0.05	<0.05

7.6 Summary of Section 7

1   Catalysts may be classified into groups depending on the types of chemical reaction for which they are active. Three important classes of catalysts are metals, metal oxides (including SiO_2) and acids (including zeolites).

2   For most metals, the strength of adsorption of simple gases and vapours decreases in the following sequence: $O_2 > C_2H_2 > C_2H_4 > CO > H_2 > N_2$.

3   Metals can be divided into groups according to their ability to adsorb simple molecules: high chemical adsorption ability is associated with transition metals.

4   The concepts underlying the idea of a volcano curve are useful in the consideration of the catalytic activity of transition metals. An acceptable measure of the strength of adsorption is the initial differential heat of chemical adsorption of an adsorbate on a polycrystalline substrate.

5   For hydrogenation and hydrogenolysis reactions the metals Ru, Rh, Pd, Os, Ir and Pt are invariably the most active. These 'platinum group' metals (plus silver) are also the most efficient *metal* catalysts for oxidation reactions.

6   Metal and semi-metal oxides of pre-transition elements are good dehydration catalysts.

7   Metal oxides of transition and post-transition elements – the semiconducting oxides – excel as oxidation catalysts. Pure oxides catalyse 'deep' oxidation reactions: mixed oxides are excellent and highly selective partial oxidation catalysts.

8   The behaviour of many partial oxidation catalysts can be rationalized in terms of a redox mechanism in which lattice oxygen is the oxidizing agent and oxygen from the gas phase serves to replenish catalyst oxygen vacancies.

9   For pure metal oxides, semiconductor type (p or n) can be used as a guide to catalytic activity: p-type oxides are usually the better oxidation catalysts.

10   The acidic character of a solid arises from the chemical mixing of two metal oxides, for example Al_2O_3 and SiO_2, which contain elements (Al and Si) of different valencies. Two types of acidic surface site can be envisaged: one due to the presence of H^+, the other a Lewis-type site. The relative importance of these sites under reaction conditions is still uncertain.

11   Zeolites are three-dimensional aluminosilicate crystals containing an internal pore structure – interconnected cavities and/or channels – of molecular dimensions.

12   Zeolites possess inherent catalytic activity, by virtue of acid sites (either Lewis or Brønsted in character) located within the pore structure and on the external surfaces. Matching the shape of reactant molecules, product molecules, and transition states with the geometry of the pore system is the basis for shape-selective catalysis by zeolites.

8 ADSORPTION ISOTHERMS

So far, our account of heterogeneous catalysis has been relatively descriptive: we must now develop a more quantitative approach. An important first step is to model the key process of chemical adsorption, because it is this process that determines the amounts of species that are adsorbed on a surface. Clearly, the 'surface concentrations' will have a marked influence on the overall kinetic behaviour of a catalysed reaction.

When a quantity of gas is admitted to an evacuated vessel containing a solid, adsorption occurs at the surface of the solid. The extent of adsorption at *equilibrium* depends on a number of factors, including the temperature, the chemical nature and surface area of the substrate, and the pressure and chemical nature of the adsorbate. Of particular interest is the relationship, for a given mass of substrate, between the amount of gas adsorbed at *constant temperature* and the pressure with which it is in equilibrium: this is called the **adsorption isotherm** (for example, Figure 34).

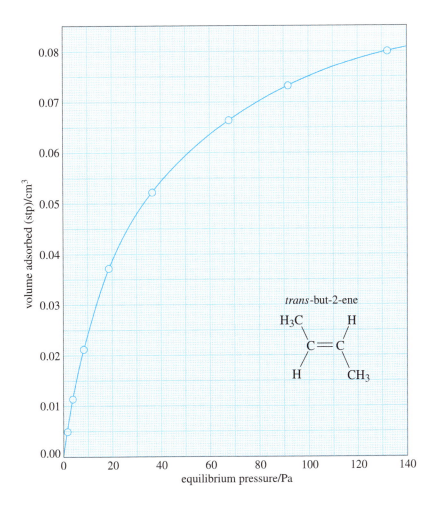

Figure 34 The adsorption isotherm for *trans*-but-2-ene on 1 g of a particular form of bismuth molybdate, Bi_2MoO_6. The isotherm was determined at 298 K, at which temperature adsorption occurs quickly and reversibly.

Note that it is common to express the amount of gas adsorbed as a volume measured in cm^3 corrected to conditions of **standard temperature and pressure (stp)**, taken in this context to be 273.15 K (0 °C) and 101.325 kPa (1 atm). The correction is straightforward. If the measured volume is V at some temperature T, and equilibrium pressure p, then the conversion to stp is:

$V(\text{stp}) = (273.15 \text{ K}/T) \times (p/101.325 \text{ kPa}) \times V$

The reason for the correction should be clear: at a fixed temperature and pressure, the volume of gas adsorbed will be directly proportional to the *number of moles* of gas adsorbed.

In general, experimental adsorption isotherms (for both chemical and physical adsorption) take a variety of forms. In this Section we shall primarily focus attention on a model first proposed by the American physical chemist, Irving Langmuir. The model has the advantage of simplicity and, as we shall see in due course, plays an important part in the analysis of mechanisms of catalysed reactions.

8.1 The Langmuir adsorption isotherm

In order to produce a working model, Langmuir envisaged the process of adsorption in a simple way by making the following assumptions:

1 Adsorbed species are attached to the surface of a substrate at definite localized sites;

2 Each adsorption site can accommodate only one adsorbed species;

3 There is the same probability of adsorption at all sites, independent of whether adjacent sites are occupied or not.

The implication of assumptions 1 and 2 is that maximum adsorption corresponds to the formation of a monolayer.

What does assumption 3 imply?

There are two implications. First, the surface of the substrate is completely uniform, so the differential heat of adsorption is independent of the fractional surface coverage. Second, the forces of attraction and repulsion between adjacent adsorbed species can be neglected.

The **Langmuir model** describes *ideal* chemical adsorption and plays a role in adsorption theory similar to that of the ideal gas equation in the theory of gases. Clearly the assumptions of the model are fairly drastic: we shall return to this criticism later in this Section, and also in Section 9 when we discuss the application of the model in kinetic studies.

To develop the model in a quantitative way, let us consider the non-dissociative adsorption of a gas, say A, on a uniform solid surface at constant temperature in a system of fixed volume. We shall also allow for the possibility of desorption. Furthermore, without loss of generality, we can assume for simplicity that the surface has unit area. The overall adsorption process can be represented by the equations:

$$A(g) + * \xrightarrow{k_a} \overset{A}{\underset{*}{|}} \qquad (32)$$

$$\overset{A}{\underset{*}{|}} \xrightarrow{k_d} A(g) + * \qquad (33)$$

in which k_a and k_d represent rate constants for adsorption and desorption, respectively. Notice that the *reversible* nature of the adsorption process has been made explicit by representing it by two separate equations and that the adsorption is at a *single* surface site.

The number of molecules of A colliding with the surface in unit time is directly proportional to the pressure of A,* that is p_A. But of course adsorption can occur only at unoccupied sites.

- Suppose that initially the surface has N adsorption sites and at some time later the fractional surface coverage is θ (see equation 18, Section 6.3.2). What is the number of unoccupied sites at this time?

- The *fraction* of unoccupied sites will be $(1 - \theta)$, so the *number* of unoccupied sites will be $N(1 - \theta)$.

- What then is the rate of adsorption of A according to equation 32?

- The rate of adsorption will be proportional to both the pressure of the gas above the surface and the number of unoccupied sites, that is:

  rate of adsorption = $k_a p_A N(1 - \theta)$ (34)

If the rate of adsorption is expressed as 'molecules' per second, then you should convince yourself that the SI unit of k_a is $Pa^{-1} s^{-1}$.

- According to equation 33, what is the rate of desorption of A?

- The rate of desorption depends only on the amount of the adsorbed gas, hence

  rate of desorption = $k_d N \theta$ (35)

Again if the rate of desorption is expressed as 'molecules' per second then the unit of k_d is s^{-1}.

---

* This is a standard result from the kinetic theory of gases; namely that the number of collisions of gas molecules with a surface of unit area in unit time is directly proportional, at a fixed temperature, to the pressure of that gas above the surface.

At equilibrium the rate of adsorption must equal that of desorption, so that equations 34 and 35 can be equated:

$$k_a p_A N(1 - \theta) = k_d N \theta \quad \text{at equilibrium} \tag{36}$$

where p_A and θ are now taken to represent *equilibrium values* of pressure and surface coverage, respectively. (Strictly, to emphasize this we should write them as $p_{A,e}$ and θ_e, respectively, but this is not the practice in the literature.) Equation 36 can be rearranged as follows:

$$k_a p_A - \theta k_a p_A = k_d \theta \tag{37}$$

so that

$$\theta = \frac{k_a p_A}{k_d + k_a p_A} \tag{38}$$

It is usual to introduce a quantity b, the so-called **adsorption coefficient**, which is defined as the ratio k_a/k_d. Dividing the top and bottom of equation 38 by k_d gives the following expression:

$$\theta = \frac{(k_a/k_d) p_A}{1 + (k_a/k_d) p_A} \tag{39}$$

then writing $b = k_a/k_d$, θ can be expressed as:

$$\theta = \frac{b p_A}{1 + b p_A} \tag{40}$$

This equation is the most widely quoted form of the **Langmuir isotherm for single-site adsorption**.

In practical terms the fractional surface coverage, θ, can be expressed as:

$$\theta = V/V_m \tag{41}$$

where V_m and V represent volumes of gas adsorbed (usually corrected to stp) for a complete monolayer and for a particular coverage, respectively.

STUDY COMMENT The following SAQ invites you to convert equation 40 into a form that can then be used to test whether an experimental isotherm fits the single-site Langmuir model. If you find difficulty with the conversion, do make sure you check our answer.

SAQ 20 Show that the equation for the Langmuir isotherm for single-site adsorption, that is equation 40, can be written in the form:

$$\frac{p_A}{V} = \frac{1}{b V_m} + \frac{p_A}{V_m} \tag{42}$$

How would you use this equation to test whether an experimental isotherm fitted the single-site Langmuir model? Some of the experimental data used to plot the adsorption isotherm shown in Figure 34 are listed in Table 5. Test whether these data fit the single-site Langmuir model. What is the value of V_m, the volume of *trans*-but-2-ene adsorbed at monolayer coverage? What is the value of the adsorption coefficient b?

Table 5 Data for the adsorption of *trans*-but-2-ene on 1 g of Bi_2MoO_6 at 298 K.

$V(stp)/cm^3$	$p(trans$-but-2-ene$)/Pa$
0.005	1.6
0.011	4.4
0.021	9.3
0.037	19.3
0.052	37.3
0.066	68.0
0.073	92.0
0.080	133.3

You may already have recognized that the adsorption coefficient b has the form of an equilibrium constant. It is, in fact, the equilibrium constant for the adsorption process, so a large value implies that the adsorption equilibrium lies well to the right and that the adsorbate is strongly adsorbed. Typically, b has the unit Pa^{-1} (as seen by dividing the unit of k_a by the unit of k_d).

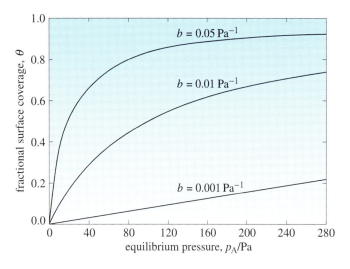

Figure 35 Langmuir isotherms for single-site adsorption calculated from equation 40.

Figure 35 compares three isotherms calculated from equation 40 with arbitrary, but realistic, values of b. One way to view the information in Figure 35 is to imagine that it depicts adsorption isotherms for three different systems at a fixed temperature. Clearly the larger the value of b, the greater the surface coverage at a given equilibrium pressure, or alternatively, the larger the value of b the lower the equilibrium pressure required to reach a particular surface coverage. But equally the figure can be viewed as representing just one adsorption system, but with the three isotherms measured at *different, fixed temperatures*. This of course implies that the magnitude of b depends on temperature, which indeed is just the type of behaviour we would expect for an equilibrium constant. In fact we can write

$$b = b_0 \exp(-\Delta H_{ad}/RT) \qquad (43)$$

where b_0 is a constant independent of temperature and ΔH_{ad} is the enthalpy change for adsorption which, in the Langmuir model, is assumed to be independent of surface coverage.

- Would you expect the magnitude of b to increase or decrease with increasing temperature for a chemical adsorption process?

- Chemical adsorption is usually exothermic, that is $\Delta H_{ad} < 0$, so the magnitude of b will *decrease* with rising temperature.

SAQ 21 The adsorption of *trans*-but-2-ene on Bi_2MoO_6 can be described by a single-site Langmuir adsorption isotherm at 323 K and at 373 K. At these temperatures values of b were determined to be 7.95×10^{-3} Pa^{-1} and 1.01×10^{-3} Pa^{-1}, respectively. What is the value of the enthalpy change for adsorption according to the Langmuir model? What can you say about the magnitude of this value?

Although in SAQ 20 you found that the experimental isotherm shown in Figure 34 fitted the single-site Langmuir model quite well, it must be emphasized that, in general, very few chemical adsorption isotherms are found to correspond to the model over their entire range of surface coverage (a state of affairs also found for physical adsorption up to monolayer coverage).

- What do you think is the main reason underlying this lack of correspondence between theory and experiment?

It is most likely a consequence of the assumption that adsorption at all sites is equally probable, whatever the extent of the surface coverage. As we saw in Section 6.3.2, for most adsorption systems the differential heat of adsorption *decreases* with increasing surface coverage, which strongly suggests that this assumption is incorrect. This then is a major criticism of the Langmuir model. As you will see in Section 9, however, the model is nonetheless useful in developing kinetic expressions for heterogeneously catalysed reactions, and there is some justification for its use in such circumstances.

Langmuir's ideas may also be extended to other cases of adsorption that are important in catalytic studies: these are briefly outlined below.

8.1.1 Dual-site adsorption

Dual-site adsorption most commonly occurs in dissociative chemisorption systems. (It can also effectively arise in some systems in which an adsorbate molecule either occupies or blocks the occupancy of a second adjacent site.) Dissociative chemisorption is taken to be a reversible process that can be represented schematically by the equation:

$$A(g) + 2* \underset{k_d'}{\overset{k_a'}{\rightleftarrows}} \begin{matrix} A' & A'' \\ | & + & | \\ * & & * \end{matrix} \qquad (44)$$

in which the gas A adsorbs as fragments A' and A'' at constant temperature. For example, if the gas were hydrogen and the substrate a transition metal, then both fragments would be hydrogen atoms (equation 11), or if the gas were propene and the substrate a bismuth molybdate-based catalyst, the fragments would be an allyl radical and a hydrogen atom (equation 12). For the purposes of analysis we shall assume that the solid surface is uniform and the system is at fixed volume.

The key to deriving the isotherm for dissociative adsorption is to recognize that, in order for adsorption to occur, a gas molecule must strike the surface at a location where there are *two adjacent* vacant sites. To obtain the number of these empty site pairs it is necessary to make some assumption about the adsorbed layer, the simplest being that the adsorbed fragments are *mobile* at all coverages. This means that, as the coverage increases, the fragments can continually rearrange themselves so that the vacant sites are always distributed randomly over the surface. The probability of any one site being empty is just the fraction of vacant sites, $(1 - \theta)$. With a random distribution, the probability of one of its neighbours being empty is *also* $(1 - \theta)$. The combined probability that *both* sites will be vacant is the product of these individual probabilities, that is $(1 - \theta)^2$. So the number of vacant pairs will be $N(1 - \theta)^2$, where N is the total number of sites on the surface. A similar argument shows that the number of *occupied* pairs will be $N\theta^2$.

■ What will be the rates of adsorption and desorption?

▨ The rate of adsorption will depend on the pressure of A, p_A, and on the number of pairs of adjacent sites, so that:

$$\text{rate of adsorption} = k_a' p_A N (1 - \theta)^2 \qquad (45)$$

Desorption will involve the interaction of two neighbouring adsorbed species, so:

$$\text{rate of desorption} = k_d' N \theta^2 \qquad (46)$$

At equilibrium the rates of adsorption and desorption are equal and so equations 45 and 46 can be equated:

$$k_a' p_A N (1 - \theta)^2 = k_d' N \theta^2 \quad \text{at equilibrium} \qquad (47)$$

where p_A and θ now represent *equilibrium values* of pressure and surface coverage, respectively. On rearrangement, equation 47 gives the **Langmuir isotherm for dissociative chemisorption**:

$$\theta = \frac{(bp_A)^{1/2}}{1 + (bp_A)^{1/2}} \qquad (48)$$

where b is the ratio k_a'/k_d'.

[Note that equation 48 represents the simplest form of isotherm for dissociative chemisorption. Strictly, it applies only if the adsorbed fragments are *mobile* at all degrees of surface coverage.]

Using the expression for θ given by equation 41, it is possible to rewrite equation 48 in the more practical form:

$$\frac{p_A^{1/2}}{V} = \frac{1}{b^{1/2} V_m} + \frac{p_A^{1/2}}{V_m} \qquad (49)$$

so that a plot of $p_A^{1/2}/V$ versus $p_A^{1/2}$ should reveal a straight line for a dissociative chemisorption system.

8.1.2 Adsorption of more than one species on the same surface

The reversible adsorption of two gases, A and B, on the same surface (at constant temperature) can be represented by the following two equations, provided it is assumed that there is only *one* type of site for adsorption and there is no dissociation:

$$A(g) + * \underset{k_d(A)}{\overset{k_a(A)}{\rightleftarrows}} \begin{array}{c} A \\ | \\ * \end{array} \quad (50)$$

$$B(g) + * \underset{k_d(B)}{\overset{k_a(B)}{\rightleftarrows}} \begin{array}{c} B \\ | \\ * \end{array} \quad (51)$$

The situation described by equations 50 and 51 is termed **competitive adsorption**. The type of analysis used in Section 8.1.1 can be used to show that, for a uniform surface, the isotherms for the equilibrium fractional surface coverages, θ_A and θ_B, are as follows:

$$\theta_A = \frac{b_A p_A}{1 + b_A p_A + b_B p_B} \quad (52)$$

$$\theta_B = \frac{b_B p_B}{1 + b_A p_A + b_B p_B} \quad (53)$$

where $b_A \; (= k_a(A)/k_d(A))$ and p_A are, respectively, the adsorption coefficient and equilibrium partial pressure of A, and $b_B \; (= k_a(B)/k_d(B))$ and p_B are, respectively, the adsorption coefficient and equilibrium partial pressure of B.

Competitive adsorption may not be limited to just two species. In the general case of the competitive adsorption of different species on a uniform surface with only one type of site, a **general Langmuir isotherm for competitive non-dissociative adsorption** can be derived:

$$\theta_A = \frac{b_A p_A}{1 + \sum_i b_i p_i} \quad (54)$$

where θ_A refers to the equilibrium fractional surface coverage of a species A, with adsorption coefficient b_A, and the summation $\sum_i b_i p_i$ is over *all* the species that adsorb (*including* A). Clearly, equations 40, 52 and 53 are just specific cases of this general equation.

Finally it is worth noting that occasionally two gases may adsorb on two *different* sets of surface sites on a given surface: this is then **non-competitive adsorption**. In this case the isotherms are just the same as they would be if adsorption occurred on two completely separate surfaces; that is, equation 40 applies separately to each gas.

8.2 Other adsorption isotherms

Various attempts have been made to develop isotherms to take into account the *non-ideal* adsorption behaviour commonly encountered in both chemical and physical adsorption systems. One such isotherm, the so-called **Temkin isotherm**, has found considerable application. The theory supposes that the fall in the differential heat of adsorption found in many chemical adsorption systems is *linearly* related to fractional surface coverage. The supposition is not unreasonable at low to medium surface coverage as evidenced, for instance, by the results displayed in Figure 14 (Section 6.3.2). One form of the isotherm is:

$$\theta = (1/c_1) \ln (c_2 b p) \quad (55)$$

where c_1 and c_2 are constants at a particular temperature, with values that depend on the magnitude of the *initial* differential heat of adsorption, b is the Langmuir adsorption coefficient, and p is the equilibrium pressure of the gas. Use of the isotherm was popularized by M. I. Temkin in his classic investigations of the catalytic decomposition and synthesis of ammonia.

Table 6 Data for the adsorption of *cis*-but-2-ene on 1 g of Bi_2MoO_6 at 298 K.

$V(stp)/cm^3$	$p(cis\text{-but-2-ene})/Pa$
0.007	1.9
0.010	4.2
0.020	20.1
0.029	54.0
0.036	102.1
0.042	170.4
0.047	261.3

STUDY COMMENT The following SAQ highlights how fitting Langmuir isotherms to experimental data can be used to help elucidate adsorption mechanisms.

SAQ 22 The adsorption of *cis*-but-2-ene on Bi_2MoO_6 occurs quickly and reversibly at 298 K. The adsorption isotherm at this temperature, determined using 1 g of Bi_2MoO_6, is shown in Figure 36, and the data used to plot the isotherm are listed in Table 6. Which of the two Langmuir isotherms, that for single-site adsorption or that for dissociative chemisorption, fits the data better? What is the value of V_m (at stp), that is the volume of adsorbed gas at full monolayer coverage?

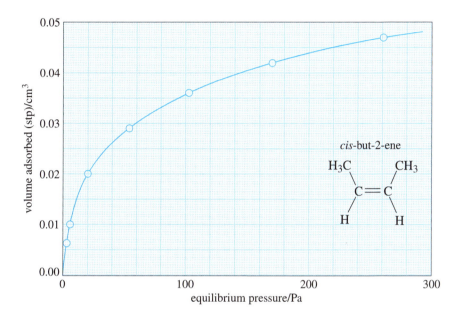

Figure 36 The adsorption isotherm for *cis*-but-2-ene on 1 g of a particular form of bismuth molybdate, Bi_2MoO_6, at 298 K.

8.3 The determination of surface area

The principal method of determining the *total* surface area of both porous and non-porous catalysts is via *physical adsorption isotherm* measurements. Because physical adsorption involves relatively non-specific forces, the same gas can be used as an adsorbate for most solids, thereby providing a 'standard' means of assessing surface area. Indeed, surface area determination is one of the few universal (or nearly so) techniques available for the characterization of solid catalysts.

In outline, the method requires the measurement of two parameters: the number, n_m, of moles of gas adsorbed in a complete, physically adsorbed monolayer, and the area occupied by a *single* adsorbed molecule in this complete monolayer; this is given the symbol Ω_m (pronounced 'omega-m'). The total surface area, S, is then given by:

$$S = \underbrace{n_m L}_{\text{number of molecules adsorbed}} \times \underbrace{\Omega_m}_{\text{area occupied by each molecule}} \quad (56)$$

where L is the Avogadro constant. It is usual to divide the total surface area by the mass of catalyst used in the determination to give the **specific surface area**: a typical unit is $m^2\,g^{-1}$.

The choice of the adsorbate gas is relatively straightforward. It should be inert and the molecules should be small and roughly spherical, so that all of the *interior* surface area of a porous solid is contacted. For these reasons nitrogen gas is mostly used (although recently the merits of argon gas have been put forward, particularly for the measurement of very small specific surface areas, $<1\,m^2\,g^{-1}$).

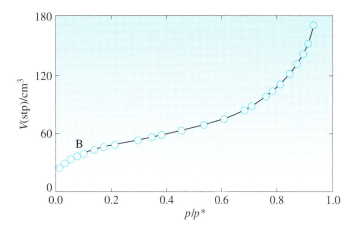

Figure 37 The physical adsorption isotherm for the adsorption of nitrogen gas on non-porous silica at 77 K.

- At what temperature would you carry out a surface area determination using nitrogen gas as an adsorbate?

- As you should recall from Section 6.2, the amount of a gas physically adsorbed always increases with decreasing temperature and is most readily determined at temperatures close to the normal boiling temperature of the adsorbate. Thus, for nitrogen gas, temperatures in the region of 77 K should be (and are) used.

The quantity n_m in equation 56 is determined from an adsorption isotherm. To see how this is accomplished we shall take as an example the adsorption of nitrogen on what has become a 'standard' solid: non-porous silica. Figure 37 shows the physical adsorption isotherm determined at 77 K: it is plotted as the volume of nitrogen adsorbed, V (expressed in cm³ at stp) against the ratio p/p^*, where p is the experimentally determined equilibrium pressure and p^* is the saturation or equilibrium vapour pressure that would exist over pure liquid nitrogen at the same temperature. (The reason for using this ratio will become apparent shortly.) The sigmoid or 'S-shaped' form of the isotherm is typical of that found for many non-porous and some porous solids. At values of the ratio p/p^* in the region of 0.1, that is at the 'knee' of the curve denoted by B in Figure 37, the amount adsorbed corresponds to a monolayer. Indeed, up to this point the isotherm is similar in character to the Langmuir single-site adsorption isotherm (cf. Figures 34 and 35). For increasing values of p/p^*, multilayer formation occurs until what is essentially a bulk liquid is present, in the limit $p/p^* = 1$.

A model for physical adsorption involving multilayer formation was first put forward by S. Brunauer, P.H. Emmett and E. Teller, in 1938: this **BET model** still reigns supreme today in surface area determinations. We shall not derive the isotherm here, but it is worth noting that, because of the assumptions it involves, it is often best regarded as a 'semi-empirical isotherm of practical utility'. The isotherm can be written in the form:

$$\frac{x}{V(1-x)} = \frac{1}{CV_m} + \frac{(C-1)x}{CV_m} \quad (57)$$

where V_m and V are the volumes of adsorbed gas (usually given as those applicable at stp) for a complete monolayer and for some coverage corresponding to an equilibrium pressure p, respectively, x is the ratio p/p^* and C is a constant that can be determined experimentally.

- Does equation 57 correspond to an equation for a straight line?

- Yes, for a plot of $x/V(1-x)$ versus x. The slope and intercept are $(C-1)/CV_m$ and $1/CV_m$, respectively.

Figure 38 shows a **BET plot** for the adsorption data in Figure 37: it is linear for the data in the region $p/p^* < 0.3$: clearly, at high p/p^* values the model cannot cope with the 'realities' of multilayer adsorption. This behaviour is typical of that found for many non-porous solids.

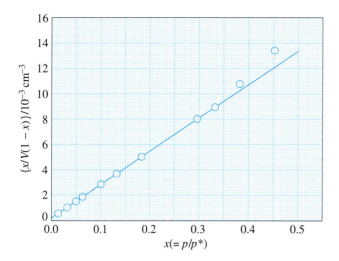

Figure 38 A BET plot for the adsorption of nitrogen on non-porous silica at 77 K.

- What is the value of V_m according to the data in Figure 38? You can assume, although you should also check this for yourself, that the slope is $2.67 \times 10^{-2}\,\text{cm}^{-3}$ and the intercept is $2.5 \times 10^{-4}\,\text{cm}^{-3}$.

- It should be clear that the value of V_m cannot be determined from the separate measurement of the slope or the intercept: but it can be determined by combining these quantities. According to equation 57:

$$\text{slope} + \text{intercept} = \frac{C-1}{CV_m} + \frac{1}{CV_m} = \frac{C-1+1}{CV_m} = \frac{1}{V_m} \qquad (58)$$

For the data above, slope + intercept = $2.695 \times 10^{-2}\,\text{cm}^{-3} = 1/V_m$. Hence

$$V_m = \frac{1}{2.695 \times 10^{-2}\,\text{cm}^{-3}} = 37.1\,\text{cm}^3 \text{ (at stp)}$$

This value of V_m can now be used to determine n_m, the number of moles of nitrogen adsorbed in a complete monolayer at 77 K.

- What is the value of n_m?

- To determine n_m, the volume V_m must be translated into moles adsorbed. *But remember this volume is expressed at stp.* Thus, assuming ideal gas behaviour:

$$n_m = \frac{pV_m}{RT} = \frac{(101\,325\,\text{Pa}) \times (37.1 \times 10^{-6}\,\text{m}^3)}{(8.314\,\text{J K}^{-1}\,\text{mol}^{-1}) \times (273.15\,\text{K})}$$
$$= 1.655 \times 10^{-3}\,\text{mol}$$

(Note carefully the units used in this calculation, and remember that $1\,\text{Pa} = 1\,\text{J m}^{-3}$.)

To complete the calculation, the effective area occupied by *each* adsorbed nitrogen molecule at 77 K, Ω_m, is required. This value is now widely accepted to be $0.162\,\text{nm}^2$ ($= 0.162\,(\times 10^{-9}\,\text{m})^2 = 0.162 \times 10^{-18}\,\text{m}^2$).

- What is the total surface area of the non-porous silica?

- According to equation 56,

$$S = (1.655 \times 10^{-3}\,\text{mol}) \times (6.022 \times 10^{23}\,\text{mol}^{-1}) \times (0.162 \times 10^{-18}\,\text{m}^2)$$
$$= 161.5\,\text{m}^2$$

- If 1 g of non-porous silica was used in the determination, what is the specific surface area?

- The specific surface area is $161.5\,\text{m}^2/1\,\text{g}$, that is $161.5\,\text{m}^2\,\text{g}^{-1}$.

Of course, many catalysts are porous. However, in many cases their nitrogen adsorption isotherms are very similar to those for non-porous solids in the region $p/p^* < 0.3$, so their surface areas can be determined using the BET method. At higher values of p/p^* the isotherms for porous solids often show a *hysteresis loop*. At the temperatures used for surface area determination, physical adsorption is completely reversible and so lowering the pressure will cause desorption. The isotherm for a non-porous solid will then simply 'retrace' its path, but on a porous solid evaporation of condensed gas in fine pores does not occur as easily as its condensation. So over the p/p^* range where the condensed material is being removed, the isotherm follows a 'higher' path than during adsorption; a schematic example is shown in Figure 39. Analysis of this type of behaviour proves to be a very valuable method of obtaining information about pore size distributions for porous catalysts.

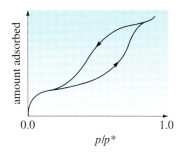

Figure 39 A schematic isotherm for the adsorption of nitrogen on a porous material at 77 K, showing a hysteresis loop.

It is important to note that the BET method measures a *total* surface area, whereas with a supported metal catalyst it is frequently desirable to measure only the area of exposed metal. This can be achieved by using an adsorbate, often hydrogen gas or carbon monoxide, that selectively chemisorbs on the metal but not on the support. If the volume adsorbed at monolayer coverage can be determined experimentally, then the metal area can be calculated.

> **STUDY COMMENT** The following SAQ describes a common method of reducing the experimental work required to carry out a surface area measurement based on the BET equation. Do make sure you work through this example and, in particular, pay special attention to the units that are involved in the calculations.

SAQ 23 In many cases of nitrogen adsorption at 77 K it is found that the slope of a linear BET plot is much larger than the intercept, the latter being close to zero. In such circumstances, for routine measurements of surface area, a 'single-point' method is often adopted: the volume of nitrogen adsorbed (corrected to stp) is measured at a value of p/p^* between 0.2 and 0.3 and it is assumed that the BET plot will pass through the origin.

For an iron catalyst promoted with alumina, the volume of nitrogen adsorbed at 77 K when p/p^* was 0.25 was determined to be 140.4 cm³ (corrected to stp). The amount of catalyst used in the determination was 49.8 g. What is the specific surface area of this catalyst?

8.4 Summary of Section 8

1   An adsorption isotherm describes the relationship between the amount of gas adsorbed at constant temperature and the pressure with which it is in equilibrium for a given amount of substrate. It is common practice to express the amount of gas adsorbed as a volume corrected to stp.

2   Various types of isotherm, derived on the basis of the Langmuir model of adsorption, are summarized in Box 1.

3   The Temkin isotherm is derived on the basis that the fall in the differential heat of adsorption found in many chemical adsorption systems is linearly related to the fractional surface coverage.

4   Surface areas of porous and non-porous catalysts can be measured using the BET model of multilayer physical adsorption. Nitrogen gas at 77 K is the most widely used adsorbate for this type of measurement, but argon may be preferred.

Box 1 Langmuir isotherms

Reaction	Description	Isotherm
$A(g) + * \rightleftharpoons \underset{*}{A\vert}$	single-site adsorption	$\theta = \dfrac{bp_A}{1+bp_A}$
$A(g) + 2* \rightleftharpoons \underset{*}{A'\vert} + \underset{*}{A''\vert}$	dissociative chemisorption	$\theta = \dfrac{(b'p_A)^{1/2}}{1+(b'p_A)^{1/2}}$
$A(g) + B(g) + 2* \rightleftharpoons \underset{*}{A\vert} + \underset{*}{B\vert}$	competitive non-dissociative adsorption	$\begin{cases}\theta_A = \dfrac{b_A p_A}{1+b_A p_A + b_B p_B}\\ \theta_B = \dfrac{b_B p_B}{1+b_A p_A + b_B p_B}\end{cases}$
$A(g) + B(g) + \ldots + i* \rightleftharpoons \underset{*}{A\vert} + \underset{*}{B\vert} + \ldots$	general competitive non-dissociative adsorption	$\theta_A = \dfrac{b_A p_A}{1+\sum_i b_i p_i}$

9 KINETICS OF SOLID-CATALYSED REACTIONS

The kinetics of reactions catalysed by solids may be studied for a variety of reasons. From the viewpoint of a chemical engineer, a knowledge of chemical kinetics is important in designing and modelling the behaviour of an industrial reactor. At a more fundamental level, measurements of rates of reaction in the presence of different catalysts can be used to correlate and, perhaps, rationalize catalytic activity; for instance, the trends in Figure 21 (Section 7.2). Finally, but by no means least, kinetic studies provide the most 'conventional' method of obtaining information about the detailed *mechanism* of solid-catalysed reactions. It is mechanism that we focus attention on in this Section.

There are several reasons why the kinetics of heterogeneous catalysis is a more difficult area of study than the kinetics of homogeneous systems. First, in heterogeneous catalysis the chemical reaction occurs on the surface of the catalyst rather than throughout the entire volume of the reaction vessel as in a homogeneous system. Second, the rate of reaction will depend on the amounts of species in the *adsorbed* layer, but in general these amounts cannot be directly measured or observed: instead they have to be indirectly related to bulk-phase pressures or concentrations. Third, the rate of reaction will depend, among other variables, on the extent of the active surface area that the catalyst can present to the reactants. Finally, heterogeneously catalysed reactions, particularly those of practical interest, invariably give rise to a range of products and thus cannot always be described by a single stoichiometric equation. Despite these difficulties, however, we shall find that many of the kinetic principles we have examined in previous Blocks can be used with only minor modification in heterogeneous catalysis.

9.1 Rate of reaction, order, and effect of temperature

The first step, as in any kinetic study, is to define the rate of reaction. It is common in heterogeneous catalysis to denote this rate by the symbol r, *but depending on the circumstances it may be defined in various ways.*

Consider a catalysed reaction that can be represented by the equation:

$$a\text{A} + b\text{B} \xrightarrow{\text{catalyst}} c\text{C} \tag{59}$$

This equation may represent the overall stoichiometry of the reaction, or it may represent just the formation of a major product in a fairly complex scheme of events. One very general method of defining the rate of reaction is as follows:

$$r = -\frac{1}{a}\frac{dn_A}{dt} = -\frac{1}{b}\frac{dn_B}{dt} = \frac{1}{c}\frac{dn_C}{dt} \tag{60}$$

where n_A, n_B and n_C are the amounts of reactants and product, respectively. Expressed in this form the rate of reaction typically has a unit of mol s^{-1}. This definition is particularly useful when dealing with catalytic reactions carried out in flow systems, which is often the method of choice for many experimental studies.

As you saw in Section 4, the rate of a heterogeneously catalysed reaction depends, in most circumstances, on the extent of the surface area of the catalyst exposed to the reactants. The ideal way of expressing this behaviour is in terms of the **turnover number**, that is the number of molecules converted per unit time per active centre. However, except in the most precise studies, which invariably involve supported metal catalysts, the number of active sites cannot usually be determined. If the total surface area, S, of the catalyst has been determined, a rate of reaction per unit area, r_s, can be quoted (it has been recommended that r_s should be termed an *areal rate of reaction*). Alternatively, if only the mass, m, of the catalyst is known, then a rate of reaction per unit mass, r_m, can be found. It should be clear that

$$r = Sr_s = mr_m \tag{61}$$

If a heterogeneously catalysed reaction is studied in a *constant volume* system, then pressure measurements are a convenient means of following the progress of the reaction. In these circumstances, for a general reaction as given in equation 59, the rate of reaction can be defined as follows:

$$r = -\frac{1}{a}\frac{dp_A}{dt} = -\frac{1}{b}\frac{dp_B}{dt} = \frac{1}{c}\frac{dp_C}{dt} \tag{62}$$

where p_A, p_B, and p_C are the partial pressures of reactants and product, respectively. Expressed in this form the rate of reaction typically has a unit of Pa s^{-1}. We shall restrict our discussion in the remainder of this Block to this particular form of definition.

Many surface-catalysed reactions have *experimental* rate equations for which an overall order of reaction can be defined over a reasonably wide temperature range. For a catalysed reaction of the form of equation 59 and occurring in a system of fixed volume, the rate equation often takes the form:

$$r = -\frac{1}{a}\frac{dp_A}{dt} = k_R\, p_A^{\alpha} p_B^{\beta} \tag{63}$$

where p_A and p_B are the partial pressures of reactants A and B, respectively, the exponents α and β are the partial orders of reaction, and k_R is the rate constant.

■ The hydrogenation of ethene, C_2H_4, to ethane, C_2H_6, over a copper catalyst,

$$C_2H_4(g) + H_2(g) \xrightarrow{\text{copper}} C_2H_6(g) \tag{64}$$

under certain conditions at 273 K has an experimental rate equation of the form

$$r = -\frac{dp_{C_2H_4}}{dt} = k_R\, \frac{p_{H_2}}{p_{C_2H_4}} \tag{65}$$

What are the partial orders of reaction with respect to each reactant, and what is the overall order of reaction?

- With respect to ethene the partial order is –1, and with respect to hydrogen it is +1. The overall order of reaction is –1 + 1 = 0, that is zero order.

There are several features worth noting about the rate equation given by equation 63. First, it turns out that partial orders of reaction are often not simple integers (or fractions such as $\frac{1}{2}$): this arises because in reality the rate equation is complex, but it can be 'forced' to fit a simple expression such as equation 63. Second, contributions from terms involving other species in the reaction mixture, such as reaction products and catalytic poisons, often occur. And finally it is important to realize that the magnitude of the rate constant k_R will depend on the particular method used for defining the rate of reaction, r; consequently, care must be exercised when using quoted values of rate constants for heterogeneously catalysed reactions.

As might be expected, rate constants for heterogeneously catalysed reactions depend on temperature: for many reactions this dependence can be described (as for homogeneous systems) over a reasonable range of temperature by an Arrhenius-type expression:

$$k_R = A_{cat} \exp(-E_{cat}/RT) \tag{66}$$

where E_{cat} is an **activation energy for the catalysed reaction** and A_{cat} is a pre-exponential factor. The complexity of the steps involved in most catalysed reactions means that these quantities are best treated as empirical parameters. The activation energy, because it occurs in an exponential term, is the predominant factor in determining the magnitude of the rate constant, and hence the rate: the smaller its value the more rapid will be the conversion of reactants into products at a given temperature. The pre-exponential factor is found experimentally to depend on both the physical form and the chemical nature of the catalyst. For instance, increasing the surface area of a catalyst, and hence the number of active sites, will enhance the rate of reaction because the magnitude of A_{cat} is effectively increased. Generally speaking, however, changes in the pre-exponential factor are not as significant as changes in activation energy in determining the rate of a catalysed reaction.

For simple catalysed reactions the magnitude of A_{cat} can be calculated, providing assumptions are made concerning the mechanisms involved. For example, if it is assumed that the rate-limiting step is the surface reaction between two adsorbed species, then it is predicted that A_{cat} will be smaller by a factor of 10^{-14} (or so) than for a comparable, bimolecular reaction occurring homogeneously in the gas phase. This in turn, as you should be able to calculate for yourself, means that at 300 K, say, the catalysed reaction must have an activation energy some 80 kJ mol^{-1} *smaller* than for the homogeneous reaction if their rates are at least to be equal. This leads to a very important generalization:

> To be efficient, a heterogeneous catalyst must provide an alternative path to reaction that has an overall activation energy considerably less than that for the corresponding homogeneous reaction (cf. Figure 2).

It is worth noting that it is quite often found that the Arrhenius parameters for a given reaction in the presence of a series of catalysts change in a particular manner: either *both* E_{cat} and A_{cat} increase or they *both* decrease. The series of catalysts can be chemically different, for example different transition metals, or they can be different preparations of the same catalyst.

- Can you see what the effect of this is? (Consider the case when both E_{cat} and A_{cat} increase.)

- An increase in E_{cat} by itself would cause a decrease in the experimental rate constant, whereas an increase in A_{cat} by itself would cause an increase. When both quantities change together, however, the change in the experimental rate constant is not so great because a change in one is 'compensated' for, in whole or in part, by a change in the other.

This so-called **compensation effect** has attracted considerable research effort, but as yet no universally applicable explanation has been forthcoming: one possible explanation links the effect with the number of active sites available, although space does not permit us to pursue the arguments here.

We should note here that the same pitfalls (inadequate temperature control, impure reactants, etc.) that frustrate investigations of homogeneous kinetics are also found in heterogeneous kinetics. But there is an additional, major danger. If *chemical information* is to be obtained from kinetic studies, it is essential to ensure that the reaction under study is neither diffusion-limited nor affected by temperature gradients (which frequently accompany concentration gradients). In practice, reactions are often studied in '*gradientless reactors*', and quantitative tests are used to determine the extent to which mass transfer effects are rate-limiting. In our discussion of mechanism in the next two Sections we shall always assume that the effects of mass or heat transfer to, or within, a catalyst can be neglected.

SAQ 24 The hydrogenolysis of ethane over a nickel catalyst supported on silica gives methane as the main product:

$$C_2H_6(g) + H_2(g) \xrightarrow{\text{nickel}} 2CH_4(g) \tag{67}$$

The reaction was studied in such a way that *initial* rate measurements could be made for a series of experiments involving different initial partial pressures of reactants at a fixed temperature. Further, the initial rates of reaction were all expressed relative to the initial rate measured under a 'standard set of conditions': in particular, an initial partial pressure of hydrogen of 0.2 atm was used (1 atm = 101.325 kPa). The results are summarized in Table 7 for a series of experiments at 464 K. Assuming that the experimental rate equation has the form:

$$-dp_{C_2H_6}/dt = k_R p_{C_2H_6}{}^\alpha p_{H_2}{}^\beta \tag{68}$$

where $p_{C_2H_6}$ and p_{H_2} represent partial pressures of ethane and hydrogen, respectively, use the information in Table 7 to determine *by inspection* the values of the partial orders of reaction α and β.

Table 7 Data for the hydrogenolysis of ethane over a supported nickel catalyst at 464 K.

$p_{C_2H_6}$ atm	p_{H_2} atm	Relative rate
0.03	0.1	3.94
0.03	0.2	1.00
0.03	0.3	0.45
0.03	0.4	0.24
0.01	0.2	0.31
0.10	0.2	3.20

9.2 Kinetic mechanisms

STUDY COMMENT This Section covers an important but conceptually quite demanding area of study. We suggest that you take time to work carefully through the examples in the text and in the SAQs.

9.2.1 Langmuir–Hinshelwood models

The first models for the kinetics of heterogeneously catalysed reactions, which still form the basis of much of present-day discussion, were proposed by I. Langmuir and C. N. Hinshelwood in the 1920s. The models, though applicable to a variety of situations, incorporate three basic assumptions: (a) the process of chemical adsorption for a reactant is taken to be at equilibrium during the course of the reaction: that is, adsorption is a fast, reversible process; (b) the rate-limiting step is the *surface* chemical reaction; (c) the underlying assumptions of the Langmuir model of adsorption are retained, so that the concentration of an adsorbed species is determined by the appropriate Langmuir isotherm.

Reactions involving a single reactant

For a reaction involving the rearrangement, or decomposition, of a single reactant, the Langmuir–Hinshelwood mechanism can be written schematically as:

adsorption: $A(g) + * \rightleftharpoons \overset{A}{\underset{*}{|}}$

surface reaction: $\overset{A}{\underset{*}{|}} \longrightarrow$ adsorbed product(s)

desorption: adsorbed product(s) $\rightleftharpoons$ free product(s)

The *overall* rate of reaction is given by that of the rate-limiting step, which the model dictates is a *unimolecular* surface step; it can be expressed as:

$$r = -dp_A/dt = k_\theta \theta_A \qquad (69)$$

where the concentration of adsorbed reactant is taken to be proportional to its equilibrium fractional surface coverage, θ_A, and k_θ can be regarded as a particular form of rate constant.

Reactions involving two reactants

For a reaction involving two gaseous reactants A and B adsorbed at adjacent sites on a catalytic surface, the Langmuir–Hinshelwood mechanism can be written schematically as:

adsorption: $A(g) + * \rightleftharpoons \overset{A}{\underset{*}{|}}$ $\qquad B(g) + * \rightleftharpoons \overset{B}{\underset{*}{|}}$

surface reaction: $\overset{A}{\underset{*}{|}} + \overset{B}{\underset{*}{|}} \longrightarrow$ adsorbed product(s)

desorption: adsorbed product(s) $\rightleftharpoons$ free product(s)

In this case the overall rate of reaction is given by that of the rate-limiting, *bimolecular* surface step and can be expressed as:

$$r = -dp_A/dt = -dp_B/dt = k_\theta \theta_A \theta_B \qquad (70)$$

where θ_A and θ_B represent the equilibrium fractional surface coverages of A and B, respectively, and k_θ is again a particular form of rate constant.

The form, and complexity, of the rate equations predicted by the schematic Langmuir–Hinshelwood models depend very much on the assumptions made concerning each specific reaction. To see why this is the case, let us examine how these models apply to experimental studies of catalysed reactions.

Table 8 (overleaf) gives the form of the experimentally determined rate equations for the decomposition of phosphine, PH_3, (equation 71) for different ranges of the initial pressure of phosphine: in all cases the temperature was in the range 880–990 K and the catalyst was tungsten.

$$PH_3(g) \xrightarrow{\text{tungsten}} P(g) + \tfrac{3}{2}H_2(g) \qquad (71)$$

Table 8 The thermal decomposition of phosphine, PH_3, in the presence of a tungsten catalyst in the temperature range 880–990 K.

Initial pressure/Pa	Experimental rate equation[a]
0.1–1.0	$r = -\dfrac{dp_{PH_3}}{dt} = k_R p_{PH_3}$
30	$r = -\dfrac{dp_{PH_3}}{dt} = \dfrac{k_R p_{PH_3}}{1 + a p_{PH_3}}$
130–670	$r = -\dfrac{dp_{PH_3}}{dt} = k_R$

[a] In the rate equation determined for an initial pressure of phosphine of 30 Pa, the quantity a is a constant for the particular reaction conditions used.

How can we account for the results in Table 8?

The first, and obvious, step is to realize that we are dealing with the decomposition of a *single* reactant. Further, *if we assume that the products of the surface decomposition are desorbed as quickly as they are formed*, that is they do not occupy any active sites on the tungsten surface, then the Langmuir isotherm for single-site adsorption of phosphine is of the same form as that given in equation 40:

$$\theta_{PH_3} = \frac{b p_{PH_3}}{1 + b p_{PH_3}} \quad (72)$$

where b is the adsorption coefficient for phosphine. Hence, according to equation 69, the overall rate of reaction is:

$$r = -\frac{dp_{PH_3}}{dt} = \frac{k_\theta b p_{PH_3}}{1 + b p_{PH_3}} \quad (73)$$

Clearly, this is of the same form (if we put $k_\theta b = k_R$ and $b = a$) as the experimentally determined rate equation in the region of initial pressures of phosphine of 30 Pa.

At *low* initial pressures of phosphine, that is when the catalytic surface is sparsely covered, the experimental rate equation is first order. Looking at equation 73, this suggests that the condition $b p_{PH_3} \ll 1$ is fulfilled, so that

$$r = -dp_{PH_3}/dt = k_\theta b p_{PH_3} \quad (74)$$

which (with $k_\theta b = k_R$) is of the same form as in Table 8.

At *high* initial pressures of phosphine, that is when the catalytic surface is almost saturated, the experimental rate equation is zero order. Looking at equation 73 in this case suggests that the opposite condition $b p_{PH_3} \gg 1$ is satisfied, so that

$$r = -dp_{PH_3}/dt = k_\theta \quad (75)$$

which (with $k_\theta = k_R$) is of the same form as in Table 8.

To summarize In general, for catalytic reactions involving only a single reactant *in which the products are weakly adsorbed*, the Langmuir–Hinshelwood mechanism predicts that the observed kinetics will gradually change from first order to zero order as the initial pressure of the reactant is increased. *Of course, for the same reaction, but using a different catalyst, the adsorption coefficient of the reactant will be different, and so the changeover to zero-order kinetics may occur in a different pressure region.* For example, the decomposition of phosphine over a molybdenum catalyst in the same temperature range as in Table 8 is already zero order at an initial pressure of phosphine of 30 Pa.

So far, we have considered only reactions in which the products are weakly adsorbed, but there is no reason why this should always be the case. For instance, the decomposition of nitrous oxide, N_2O, over platinum metal in the temperature range 870–1 470 K,

$$N_2O(g) \xrightarrow{\text{platinum}} N_2(g) + \tfrac{1}{2}O_2(g) \tag{76}$$

produces oxygen, which is strongly adsorbed (cf. Section 7.2). For relatively low initial pressures of N_2O, the experimental rate equation takes the form:

$$r = -\frac{dp_{N_2O}}{dt} = \frac{k_R p_{N_2O}}{1 + a p_{O_2}} \tag{77}$$

where a is a constant for a given set of experimental conditions. The reaction is thus *inhibited* by the product oxygen.

■ How can we explain the form of this rate equation in terms of a Langmuir–Hinshelwood mechanism for a reaction containing a single reactant?

To start the analysis, assume that the rate of the decomposition reaction depends on the amount of N_2O adsorbed on the surface, then (from equation 69):

$$r = -dp_{N_2O}/dt = k_\theta \theta_{N_2O} \tag{78}$$

■ What is the form of the Langmuir isotherm for N_2O in the presence of O_2?

■ For the competitive adsorption of N_2O and O_2 according to equation 52,

$$\theta_{N_2O} = \frac{b_{N_2O} p_{N_2O}}{1 + b_{N_2O} p_{N_2O} + b_{O_2} p_{O_2}} \tag{79}$$

Because oxygen is strongly adsorbed, b_{O_2} will be much greater than b_{N_2O} (cf. Section 8.1) and so, for *low* pressures of N_2O, it is reasonable to assume that $1 + b_{O_2} p_{O_2}$ is much greater than $b_{N_2O} p_{N_2O}$. Equation 79 can then be simplified:

$$\theta_{N_2O} = \frac{b_{N_2O} p_{N_2O}}{1 + b_{O_2} p_{O_2}} \tag{80}$$

Hence, combining equations 78 and 80 gives a *theoretical* rate equation,

$$r = -\frac{dp_{N_2O}}{dt} = \frac{k_\theta b_{N_2O} p_{N_2O}}{1 + b_{O_2} p_{O_2}} \tag{81}$$

which is of the same form (with $k_R = k_\theta b_{N_2O}$ and $a = b_{O_2}$) as that found experimentally.

To generalize The method described above can be extended to consider the case of inhibition by more than one product, and to inhibition caused by the presence of non-reacting substances, or catalyst poisons, which compete for active sites on the catalyst surface.

The treatment of catalysed reactions involving two reactants using a Langmuir–Hinshelwood approach is very similar to that already developed for a single reactant. As might be anticipated, however, the equations are more complex since the number of possibilities for strong, moderate and weak adsorption is increased. With this latter point in mind, it is useful to recall the definition of the general Langmuir isotherm for (non-dissociative) competitive adsorption:

$$\theta_A = \frac{b_A p_A}{1 + \sum_i b_i p_i} \tag{54}$$

where θ_A refers to the equilibrium fractional surface coverage of a particular species A with adsorption coefficient b_A, and the summation $\sum_i b_i p_i$ is over *all* the species that adsorb (including A).

■ What is the most general form of the Langmuir–Hinshelwood rate equation for a catalysed reaction of the form:

$$A(g) + B(g) \xrightarrow{\text{catalyst}} \text{products} \tag{82}$$

The overall rate of reaction is given by equation 70:

$$r = -dp_A/dt = -dp_B/dt = k_\theta \theta_A \theta_B \qquad (70)$$

Substituting for *both* θ_A and θ_B in this expression (using equation 54) yields:

$$r = -\frac{dp_A}{dt} = -\frac{dp_B}{dt} = \frac{k_\theta b_A b_B p_A p_B}{\left(1 + \sum_i b_i p_i\right)^2} \qquad (83)$$

STUDY COMMENT In general, if you are asked to relate the expression in equation 83 to an experimental rate equation, there are two major limiting conditions (of several) that can be applied. These are described below.

1 If *all* the species are weakly adsorbed (i.e. all the b_i are small) and/or *all* the pressures, p_i, are low, then it is possible that $\sum_i b_i p_i$ will be much less than 1, and the expression in the denominator of equation 83 will reduce to 1. Under these circumstances, equation 83 will reduce to the following simplified form:

$$r = \frac{k_\theta b_A b_B p_A p_B}{(1)^2} = k_R p_A p_B \qquad (84)$$

where $k_R = k_\theta b_A b_B$.

2 If *one* species (call it X, it may be a reactant or product) is strongly adsorbed and/or p_X is high, while the remaining species are only weakly adsorbed and/or are at low pressure, then it is possible that the term $b_X p_X$ will dominate the expression in the denominator of equation 83. Under these circumstances, the equation will reduce to the following form:

$$r = \frac{k_\theta b_A b_B p_A p_B}{(1 + b_X p_X)^2} \qquad (85)$$

If the experimental conditions are such that $b_X p_X \gg 1$ (i.e. if X is very strongly adsorbed and/or at high pressure), then this equation reduces to:

$$r = \frac{k_\theta b_A b_B p_A p_B}{(b_X p_X)^2} \qquad (86)$$

Finally, if X is one of the reactants (A or B), then this expression can be further simplified; for example, if X is the reactant B then,

$$r = k_\theta \frac{b_A p_A}{b_B p_B} \qquad (87)$$

STUDY COMMENT You should now try applying these limiting conditions by working through the following SAQ.

SAQ 25 The hydrogenation of ethene to ethane, over a copper catalyst

$$C_2H_4(g) + H_2(g) \xrightarrow{\text{copper}} C_2H_6(g) \qquad (64)$$

has been studied in the temperature range 273–473 K. At 273 K the experimental rate equation is

$$r = -dp_{C_2H_4}/dt = k_R \frac{p_{H_2}}{p_{C_2H_4}} \qquad (65)$$

whereas at 473 K it takes the form

$$r = -dp_{C_2H_4}/dt = k_R p_{C_2H_4} p_{H_2} \qquad (88)$$

Assuming that there is no dissociation of either the reactants or the product on the copper catalyst, see if you can rationalize these results in terms of a Langmuir–Hinshelwood mechanism for the reaction.

For catalytic reactions involving two reactants there is also the possibility that reaction may occur between a chemisorbed species and a molecule reacting with it directly, either from the gas phase or from a physically adsorbed layer: in either case a new chemisorbed species will be formed at the surface. If this is taken as the rate-limiting step then for a general reaction, such as that given by equation 82, in which species B is adsorbed, the overall rate of reaction is given by:

$$r = -dp_A/dt = -dp_B/dt = k_\theta p_A \theta_B \tag{89}$$

This idea was first put forward by Langmuir in 1921 and later revised by E. K. Rideal in 1939 and D. D. Eley and E. K. Rideal in 1941: for this reason such a mechanism is variously referred to as the **Rideal or Eley–Rideal mechanism**. In practice there are few unambiguous examples of this type of mechanism, but the form of the theoretical rate equation can be predicted by making suitable substitutions for θ_B in equation 89.

Langmuir–Hinshelwood (or Rideal) mechanisms can be used to predict the temperature behaviour of catalysed reactions. If we return to the basic equations for the overall rates of surface reactions, that is equations 69 and 70, then it is reasonable to suppose that the rate constant k_θ will depend on temperature in an Arrhenius way. However, it must be remembered that the equilibrium fractional surface coverages θ_A and θ_B are also temperature dependent. This 'additional' temperature dependence will be most evident when reactions are studied over an extended temperature range: it may be used to explain the change in form of an experimental rate equation (as in SAQ 25 for example), or it may be used to account for the non-Arrhenius behaviour of an experimentally determined rate constant as a function of temperature.

Finally, we must consider the limitations of the Langmuir–Hinshelwood approach to describing the mechanism of heterogeneously catalysed reactions.

Can you suggest what these are?

One fundamental limitation is the assumption that the rate-determining step is an irreversible surface reaction. More modern approaches, though still very much based on the original ideas of Langmuir and Hinshelwood, take all three steps, that is adsorption, surface reaction and desorption, to be reversible and then consider the rate equations that arise *if any one* of these processes is rate-limiting. Tabulation of theoretical rate equations for a wide variety of general reaction schemes is then possible. Such an approach was pioneered by O. A. Hougen and K. M. Watson, and lists of rate expressions are often referred to collectively by their names, especially in the chemical engineering literature.

Another limitation is the implicit assumption that all surface reactions involve just a single unimolecular, or bimolecular, surface step. It is more than likely that 'real' reactions involve the formation and disappearance of a number of adsorbed species on the catalytic surface. The detection of such species *during* a reaction would, however, be an enormously difficult task. Indeed, even with modern techniques, it is true to say that a detailed mechanism for even the 'simplest' catalytic reaction is difficult to determine with certainty.

Yet another limitation lies in adopting the Langmuir model for adsorption: is this realistic when most experimental adsorption studies show that ideal adsorption behaviour is rarely found? This criticism is difficult to answer in detail, but it is possible to argue that it may not be too serious. Consider the following proposal. Molecules adsorbed on the most active sites may be too firmly bound to take part in surface reactions. Consequently, it is species of intermediate binding strength that participate in the catalytic reaction. Such species will be adsorbed over a limited range of surface coverage, and so the Langmuir model may well apply in the sense that the differential binding energy will be roughly constant.

Despite these limitations, Langmuir–Hinshelwood mechanisms are very useful in discussing the kinetics of catalysed reactions. They may suggest how the reaction conditions, or the composition of the reaction mixture, may be changed to increase catalytic activity and selectivity, and they provide a framework for discussing self-inhibition and catalytic poisoning. Simple mechanisms for laboratory reactions can also provide a starting point for the empirical description of reactions on the industrial scale. The relative simplicity of the mechanisms also fulfils another 'guiding rule' in heterogeneous catalysis. Because little is known about the details of adsorbed species and their reactions on surfaces, any detailed mechanistic formulation would contain a large number of uncertain parameters with a consequent decline in predictive capabilities. A guiding rule is, then, to provide models that, although realistic, are no more complicated than is necessary to explain the experimental observations. Of course (as we shall see in the next Section) a drawback of such an approach is that these models are unlikely to be unique.

9.2.2 Two-step models and the steady-state approximation

As we have already indicated, it is more than likely that most heterogeneously catalysed reactions involve several intermediate surface species and consequently have quite complex multi-step mechanisms. The question thus arises as to how to formulate the *simplest* type of rate equation that can adequately represent such multi-step mechanisms. The problem is not a new one, as shown for example by the work of Mars and van Krevelen on redox mechanisms (Section 7.3). One formal approach to the problem is to suggest that a multi-step surface reaction can usefully be treated by making two simplifying assumptions: (a) one step is rate-limiting; (b) one surface intermediate is the predominant one, that is the concentrations of all other surface reactive species are relatively small. The price to be paid for such an approach is, as we shall see shortly, the well-known ambiguity of simplified kinetics.

To illustrate the procedure let us concentrate on a specific example. The dehydrogenation of methylcyclohexane (MCH), C_7H_{14}, to toluene (T), C_7H_8, over a platinum/alumina catalyst,

$$C_7H_{14}(g) \xrightarrow{Pt/Al_2O_3} C_7H_8(g) + 3H_2(g) \tag{90}$$

was found, at least in its initial stages, to have an experimental rate equation of the form:

$$r = -\frac{dp_{MCH}}{dt} = \frac{k_R p_{MCH}}{1 + a p_{MCH}} \tag{91}$$

- Do you recognize the form of this equation?

- It is of the same form as would be predicted by a Langmuir–Hinshelwood model, where the products are weakly adsorbed (cf. equation 73 for the decomposition of phosphine, Section 9.2.1).

However, despite the similarities with a Langmuir–Hinshelwood mechanism, an alternative *two-step* mechanism for the reaction can be proposed (for clarity, we have not used the asterisk notation here):

$$C_7H_{14}(g) + \text{(adsorption site)} \xrightarrow{k_1} C_7H_8(ad) + 3H_2(g) \tag{92}$$

$$C_7H_8(ad) \xrightarrow{k_2} C_7H_8(g) + \text{(adsorption site)} \tag{93}$$

The first step, which is taken to be irreversible, implies that methylcyclohexane undergoes rapid dissociative adsorption with immediate desorption of the hydrogen formed. The second step, *which is taken to be rate-limiting*, is the irreversible desorption of toluene: all other surface intermediates are taken to be present in insignificant amounts.

- If the fractional surface coverage of toluene, *at some time in the reaction*, is θ_T, what fraction of sites are vacant?

- The fraction of sites vacant will be $(1 - \theta_T)$.

The adsorbed toluene can be treated as a reaction intermediate in just the same way as in homogeneous kinetics (cf. Block 3).

- What, therefore, is the rate of change of 'concentration' of adsorbed toluene with time?

- rate $= k_1 p_{MCH}(1 - \theta_T) - k_2 \theta_T$ \hfill (94)

In the *steady state*, the rate of change of 'concentration' of the adsorbed toluene will be effectively zero, that is the rates of steps 92 and 93 are identical, so that

$$k_1 p_{MCH}(1 - \theta_T) = k_2 \theta_T \qquad (95)$$

or, on expanding the left-hand side and collecting terms in θ_T,

$$k_1 p_{MCH} = \theta_T(k_2 + k_1 p_{MCH}) \qquad (96)$$

So

$$\theta_T = \frac{k_1 p_{MCH}}{k_2 + k_1 p_{MCH}} \qquad (97)$$

■ What, then, is the overall rate of reaction?

▫ It is just that of the rate-limiting step (equation 93):

$$r = -\frac{dp_{MCH}}{dt} = k_2 \theta_T = \frac{k_1 k_2 p_{MCH}}{k_2 + k_1 p_{MCH}} \qquad (98)$$

Simplifying the expression on the right (by dividing top and bottom by k_2), the overall rate of reaction can be written as

$$r = -\frac{dp_{MCH}}{dt} = \frac{k_1 p_{MCH}}{1 + (k_1/k_2) p_{MCH}} \qquad (99)$$

that is, in a form similar to that determined experimentally (equation 91) with $k_R = k_1$ and $a = k_1/k_2$.

The example serves to highlight the ambiguity of simple kinetic models: either a Langmuir–Hinshelwood or a two-step model would appear to explain the experimental rate equation quite adequately. But notice that the former model assumes that the methylcyclohexane is relatively strongly adsorbed and the toluene weakly adsorbed, whereas the two-step model assumes exactly the reverse! Very detailed kinetic tests are required if the two mechanisms are to be distinguished from one another.

Finally it is important to notice the use of the steady-state approximation as applied to a surface-adsorbed species in the above treatment. In fact the application is quite general in the analysis of surface mechanisms.

SAQ 26 The vapour-phase oxidation of toluene, C_7H_8, at 600 K over a commercial vanadium pentoxide catalyst promoted with potassium sulfate gives rise to a range of partially oxidized products (including benzaldehyde, benzoic acid, maleic anhydride and *p*-benzoquinone):

$$C_7H_8(g) + nO_2(g) \xrightarrow{V_2O_5/K_2SO_4} \text{products} \qquad (100)$$

The quantity n is the number of moles of oxygen used per mole of toluene oxidized; experimentally its value is found to be close to unity. Two possible mechanisms for the reaction are: (a) a Rideal mechanism in which the rate-limiting step is the reaction between adsorbed molecular oxygen, as the predominant surface species, and gas-phase toluene; (b) a two-step redox mechanism involving adsorbed molecular oxygen:

$$O_2(g) + \text{(adsorption site)} \xrightarrow{k_1} O_2(ad) \qquad (101)$$

$$O_2(ad) + C_7H_8(g) \xrightarrow{k_2} \text{products} \qquad (102)$$

in which both steps are irreversible and a steady-state concentration of adsorbed oxygen is established. Once again, adsorbed molecular oxygen is the predominant surface species. The second step can be taken to be rate-limiting.

Derive rate equations for these two mechanisms assuming $n = 1$. Can you suggest how measurements of the overall initial rate of reaction as a function of the initial partial pressure of toluene (while keeping the initial partial pressure of oxygen *fixed*) can be used to distinguish between the two mechanisms? [*Hint* Consider how the reciprocal of the initial rate of reaction, that is $1/r_0$, varies under these conditions.]

9.3 Summary of Section 9

1 Box 2 summarizes some of the main kinetic features of a heterogeneously catalysed reaction occurring in a system of *constant volume*.

Box 2

$$aA + bB \xrightarrow{\text{catalyst}} cC$$

overall rate of reaction, $r = -\dfrac{1}{a}\dfrac{dp_A}{dt} = -\dfrac{1}{b}\dfrac{dp_B}{dt} = \dfrac{1}{c}\dfrac{dp_C}{dt}$

typical rate equation: $r = k_R p_A^{\alpha} p_B^{\beta}$

where k_R is the rate constant: $k_R = A_{cat} \exp(-E_{cat}/RT)$.

---

2 The activation energy is the predominant factor in determining the magnitude of the rate constant and hence the rate of the catalysed reaction.

3 The Langmuir–Hinshelwood models for heterogeneously catalysed reactions assume that: (a) the process of chemical adsorption of a reactant is at equilibrium during a reaction; (b) the rate-limiting step is that of a single surface chemical reaction; (c) chemical adsorption can be described by appropriate Langmuir isotherms.

4 Box 3 summarizes the basic expressions for the overall rates of reaction, r, for various types of reaction according to Langmuir–Hinshelwood or Rideal models.

Box 3

Langmuir–Hinshelwood

single reactant: $r = -dp_A/dt = k_\theta \theta_A$

two reactants: $r = -dp_A/dt = -dp_B/dt = k_\theta \theta_A \theta_B$

Rideal

two reactants: $r = -dp_A/dt = -dp_B/dt = k_\theta p_A \theta_B$

The forms of theoretical rate equations can be predicted by substituting appropriately for θ_A and θ_B in these expressions. Simplifications can be made depending on the initial partial pressures of reactants and the relative strengths of adsorption of reactants or products. An outline of two important limiting conditions is given in Box 4.

Box 4

For the reaction

$$A + B \xrightarrow{\text{catalyst}} \text{products}$$

the general form of the Langmuir–Hinshelwood rate equation can be expressed as

$$r = -\frac{dp_A}{dt} = -\frac{dp_B}{dt} = \frac{k_\theta b_A b_B p_A p_B}{\left(1 + \sum_i b_i p_i\right)^2}$$

If *all* the species are weakly adsorbed and/or *all* the pressures, p_i, are low, it is possible that $1 \gg \sum_i b_i p_i$. Then:

$$r = -\frac{dp_A}{dt} = -\frac{dp_B}{dt} = k_R p_A p_B, \text{ where } k_R = k_\theta b_A b_B$$

If one species X (where X = A, B *or* a product) is strongly adsorbed, and/or p_X is high, while the remaining species are weakly adsorbed and/or at low pressure, then it is possible that $b_X p_X$ predominates in the sum $\sum_i b_i p_i$, and then:

$$r = -\frac{dp_A}{dt} = -\frac{dp_B}{dt} = \frac{k_\theta b_A b_B p_A p_B}{(1 + b_X p_X)^2}$$

If, *in addition*, the conditions are such that $b_X p_X \gg 1$ then:

$$r = -\frac{dp_A}{dt} = -\frac{dp_B}{dt} = \frac{k_\theta b_A b_B p_A p_B}{(b_X p_X)^2}$$

5   The main limitations of Langmuir–Hinshelwood mechanisms are: (a) the assumption that the rate-limiting step is always that of an irreversible surface reaction; (b) the assumption that all surface reactions involve a single elementary surface step; (c) the use of isotherms that pertain to ideal adsorption.

6   Multi-step surface reactions can often be usefully treated by making two simplifying assumptions: (a) one step is rate-limiting; (b) one surface intermediate is predominant. These simplifications lead to a 'condensed' two-step mechanism, which can be analysed by applying the steady-state approximation to the surface intermediate.

OBJECTIVES FOR BLOCK 5

Now that you have completed Block 5, you should be able to do the following things:

1   Recognize valid definitions of, and use in a correct context, the terms, concepts and principles printed in bold type in the text and collected in the following Table.

List of scientific terms, concepts and principles used in Block 5

Term	Page No	Term	Page No
acid catalyst	37	Langmuir model	46
acidity of zeolites	41	Lennard-Jones potential energy plot	20
activated chemical adsorption	22	Mars and van Krevelen mechanism	35
activation energy for chemical adsorption	22	monolayer coverage	20
active site, or active centre	9	multilayer formation	20
adsorbate	16	non-activated chemical adsorption	22
adsorption	7	non-competitive chemical adsorption	50
adsorption coefficient, b	47	physical adsorption	17
adsorption isotherm	44	poisoning	15
associative chemical adsorption	19	poison-resistant promoter	15
BET plot	52	pore structure (of zeolites)	38
catalyst activity	14	potential energy description of heterogeneous catalysis	8
catalyst selectivity	14	precursor state (for chemical adsorption)	21
catalyst stability	15	promoter	11
catalyst support	10	rate of a catalysed reaction, r	56
chemical adsorption (chemisorption)	17, 19	Rideal or Eley–Rideal mechanism	63
chemical (or electronic) promoter	12	semiconducting oxides	34
competitive chemical adsorption	50	shape-selective catalysis (by zeolites)	41
desorption	8	silica–aluminas	37
dissociative chemical adsorption	19	sinter	10
enthalpy change for adsorption, ΔH_{ad}	8, 24	structural promoter	11
enthalpy change for chemical adsorption, ΔH_{ca}	19	substrate	16
enthalpy change for physical adsorption, ΔH_{pa}	18	surface area	9
fouling	15	surface heterogeneity	27
fractional surface coverage, θ	24	thermal desorption spectroscopy	27
heat of adsorption	25	turnover number	56
Langmuir adsorption isotherm, competitive	50	two-step mechanism	64
Langmuir adsorption isotherm, dissociative	49	volcano curve	32
Langmuir adsorption isotherm, single-site	47	zeolites	38
Langmuir–Hinshelwood mechanisms	58		

2   Describe, in terms of a potential energy picture, the five steps that constitute the total process of heterogeneous catalysis, and state concisely the main features of a solid catalyst.

3   Explain the practical advantages to be gained by preparing solid catalysts in a particular physical form. (SAQ 1)

4   Indicate how solid catalysts, particularly those used in industrial reactions, may be classified according to their physical form. (SAQ 3)

5   Use thermodynamic calculations to determine the conditions – pressure, and particularly temperature – under which a catalyst must operate if reasonable conversion into products is to be achieved. (SAQs 4 and 5)

6   Describe the main factors that determine the suitability of a catalyst for an industrial or technological application. (SAQs 6, 16 and 18)

7   Explain in simple terms why molecules adsorb at solid surfaces.

8   Describe the main features of physical adsorption. (SAQs 7 and 14)

9   Describe the main features of chemical adsorption and indicate what is meant by the terms 'dissociative chemisorption' and 'associative chemisorption'. (SAQs 8 and 14)

10   Indicate how a Lennard-Jones potential energy plot can be used to describe the process of dissociative chemisorption and explain the difference between activated and non-activated adsorption processes. (SAQs 9, 10 and 12)

11   Explain why chemical adsorption is virtually always an exothermic process. (SAQ 11)

12   Explain why differential heats of adsorption depend on the fractional surface coverage. (SAQ 12)

13   Describe the technique of thermal desorption spectroscopy, and outline the type of information that the technique provides. (SAQ 13)

14   Discuss the link between chemical adsorption studies and heterogeneous catalysis.

15   Given suitable data, show how the ability of metals to chemically adsorb simple molecules can be related to their catalytic properties.

16   Outline, giving an example, a redox mechanism.

17   Explain, giving examples, why semiconducting metal oxides excel as oxidation catalysts. (SAQs 15 and 16)

18   Explain experimental data on a zeolite-catalysed reaction in terms of the acidity and/or shape selectivity of the zeolite. (SAQs 17–19)

19   State the assumptions underlying the derivation of Langmuir isotherms for various types of chemical adsorption, derive theoretical isotherms, and test whether they fit experimental data. (SAQs 20 and 22)

20   Calculate the enthalpy change on adsorption, given suitable data concerning the effect of temperature on the Langmuir adsorption coefficient. (SAQ 21)

21   Explain how the BET model of multilayer physical adsorption may be used to measure the total (or specific) surface area of a catalyst. (SAQ 23)

22   Define the terms 'rate of reaction', 'order of reaction' and 'Arrhenius parameters' as applied to a heterogeneously catalysed reaction. (SAQ 24)

23   Show how Langmuir–Hinshelwood mechanisms of heterogeneously catalysed reactions can be used to predict the form and temperature dependence of experimental rate equations, and state the limitations of these models. (SAQs 25 and 26)

24   Use the steady-state approximation in the analysis of simplified two-step mechanisms for heterogeneously catalysed reactions. (SAQ 26)

SAQ ANSWERS AND COMMENTS

SAQ 1 (Objective 3)

Let the number of crystallites be n. Then

$$\text{total surface area of crystallites} = n \times 4\pi(d/2)^2 = n\pi d^2$$

where d is the diameter of a crystallite. The mass m, of a *single* crystallite will be equal to its density times its volume:

$$m = \rho \times \tfrac{4}{3}\pi(d/2)^3 = \rho\pi d^3/6$$

If the total mass of metal is M, then

$$n = M/m = 6M/\rho\pi d^3$$

Hence the total surface area of crystallites is

$$(6M/\rho\pi d^3)\pi d^2 = 6M/\rho d$$

When $M = 1$ g, $\rho = 21.45 \times 10^6$ g m^{-3} and $d = 5 \times 10^{-9}$ m, then the total surface area of crystallites is

$$\frac{6 \times 1 \text{ g}}{(21.45 \times 10^6 \text{ g m}^{-3}) \times (5 \times 10^{-9} \text{ m})} = 55.9 \text{ m}^2$$

SAQ 2

A mass of 100 g of the palladium precursor solution will contain 8.13 g of palladium, therefore 5 g of solution contains

$$\frac{5}{100} \times 8.13 \text{ g Pd} = 0.406 \text{ g Pd}$$

Assuming that 90% of the precursor is retained, then

$$\text{mass Pd retained} = (0.406 \text{ g})\left(\frac{90}{100}\right) = 0.365 \text{ g}$$

Assuming that after activation all of the Pd retained remains on the surface, then

$$\text{Pd mass \%} = \frac{0.365 \text{ g Pd}}{10 \text{ g} + 0.365 \text{ g}} \times 100\% = 3.52\%$$

Hence the catalyst contains 3.52 mass % palladium.

SAQ 3 (Objective 4)

Table 9 classifies the industrial catalysts given in Table 1 in terms of their physical form. For each category the various examples are listed in the order in which they appear in Table 1. Supported, or unsupported, multicomponent systems predominate.

Table 9 Classification of industrial catalysts in terms of their physical form.

Physical form	Examples
unsupported metals or alloys	90% platinum/10% rhodium wire gauze, Raney nickel
very high surface area materials	amorphous silica–aluminas, zeolites
supported dispersions: binary systems	nickel on support, platinum on acidified γ-alumina, silver on support, palladium on suitable support, platinum on support, chromium oxide/alumina
supported, or unsupported multicomponent systems	vanadium(V) oxide plus potassium sulfate on silica, iron promoted with aluminium, potassium, calcium and magnesium oxides, etc. – in fact this category covers the remaining examples in Table 1

SAQ 4 (Objective 5)

The calculation is very similar to the one in part (b) of Exercise 1 at the end of Block 1; there you dealt with the synthesis of methanol from synthesis gas. We shall use the same notation as developed in that Block.

$$CH_4(g) + NH_3(g) = HCN(g) + 3H_2(g) \tag{4}$$

(a) As a first step we need to determine the value of K_p and hence $K^\ominus$, when $y = 0.2$ and $p_{tot} = 1$ bar:

$$K_p = \frac{p(HCN)\{p(H_2)\}^3}{p(CH_4)p(NH_3)}$$

We define the equilibrium yield of hydrogen cyanide as:

$$y = p(HCN)/p_{tot}$$

so we need to find the temperature at which $p(HCN) = 0.2 p_{tot}$. In addition, because $p(H_2) = 3p(HCN)$ throughout the reaction, it follows that $p(H_2) = 0.6 p_{tot}$.

In general, at equilibrium (from Dalton's law):

$$p_{tot} = p(CH_4) + p(NH_3) + p(HCN) + p(H_2)$$

$$= p(CH_4) + p(NH_3) + 0.8 p_{tot}$$

Further, the reactants were mixed in stoichiometric proportions so that $p(CH_4) = p(NH_3)$ throughout the reaction, and so

$$p(NH_3) = p(CH_4) = \frac{(1-0.8)p_{tot}}{2} = 0.1 p_{tot}$$

Hence, collecting terms:

$$K_p = \frac{(0.2 p_{tot}) \times (0.6 p_{tot})^3}{(0.1 p_{tot}) \times (0.1 p_{tot})} = 4.320 p_{tot}^2$$

$$= 4.320 \text{ bar}^2 \text{ (with } p_{tot} = 1 \text{ bar)}$$

Thus $K^\ominus = 4.320$ and $\ln K^\ominus = 1.4633$.

The temperature is calculated from the expression (Block 1)

$$T = \Delta H_m^\ominus / (\Delta S_m^\ominus - R \ln K^\ominus)$$

For reaction 4,

$$\Delta H_m^\ominus (298.15 \text{ K}) = \{135.1 + 3(0) - (-74.8) - (-46.0)\} \text{ kJ mol}^{-1}$$

$$= 255.9 \text{ kJ mol}^{-1}$$

$$\Delta S_m^\ominus (298.15 \text{ K}) = \{201.8 + (3 \times 130.7) - 186.3 - 192.5\} \text{ J K}^{-1} \text{ mol}^{-1}$$

$$= 215.1 \text{ J K}^{-1} \text{ mol}^{-1}$$

so that

$$T = \frac{255.9 \times 10^3 \text{ J mol}^{-1}}{(215.1 \text{ J K}^{-1} \text{ mol}^{-1}) - (8.314 \text{ J K}^{-1} \text{ mol}^{-1} \times 1.4633)} = 1\,261 \text{ K}$$

(b) Assumptions include: (i) that $\Delta H_m^\ominus$ and $\Delta S_m^\ominus$ are independent of temperature (this is not such a good assumption in view of the large temperature range involved); (ii) that CH_4, NH_3, HCN, H_2, and the mixture, behave ideally; (iii) that there are no other possible competing reactions.

[Notice that it is important at the high temperatures involved that the catalyst chosen for the reaction should not cause ammonia to decompose rapidly. The reaction is in fact carried out industrially (the Degussa process) at about 1 500 K using a supported platinum catalyst. However, the alternative Andrussov process

$$CH_4(g) + NH_3(g) + \tfrac{3}{2}O_2(g) = HCN(g) + 3H_2O(g)$$

is the principal present-day process for the production of hydrogen cyanide.]

SAQ 5 (Objective 5)

For reaction 5, $C_2H_5OH(g) = CH_3CHO(g) + H_2(g)$,

$\Delta H_m^\ominus (298.15 \text{ K}) = \{-166.2 + 0 - (-235.1)\} = 68.9 \text{ kJ mol}^{-1}$

$\Delta S_m^\ominus (298.15 \text{ K}) = \{250.3 + 130.7 - 282.7\} = 98.3 \text{ J K}^{-1} \text{ mol}^{-1}$

For reaction 6, $C_2H_5OH(g) = C_2H_4(g) + H_2O(g)$,

$\Delta H_m^\ominus (298.15 \text{ K}) = \{52.3 - 241.8 - (-235.1)\} = 45.6 \text{ kJ mol}^{-1}$

$\Delta S_m^\ominus (298.15 \text{ K}) = \{219.6 + 188.8 - 282.7\} = 125.7 \text{ J K}^{-1} \text{ mol}^{-1}$

Notice that both reactions are endothermic, and so the values of their respective standard equilibrium constants will increase with increasing temperature. Also both reactions have $\Delta S_m^\ominus > 0$, so in both cases the value of $\Delta G_m^\ominus$ will become more negative (or less positive) with increasing temperature.

Assuming that both $\Delta H_m^\ominus$ and $\Delta S_m^\ominus$ are temperature independent, and using the equations:

$\Delta G_m^\ominus = \Delta H_m^\ominus - T \Delta S_m^\ominus$

$\Delta G_m^\ominus = -RT \ln K^\ominus$

For reaction 5, $\Delta G_m^\ominus (600 \text{ K}) = 9.92 \text{ kJ mol}^{-1}$ and so $K^\ominus = 0.14$ at 600 K, and $\Delta G_m^\ominus (1\,000 \text{ K}) = -29.40 \text{ kJ mol}^{-1}$ and so $K^\ominus = 34.34$ at 1 000 K.

For reaction 6, $\Delta G_m^\ominus (600 \text{ K}) = -29.82 \text{ kJ mol}^{-1}$ and so $K^\ominus = 3.95 \times 10^2$ at 600 K, and $\Delta G_m^\ominus (1\,000 \text{ K}) = -80.10 \text{ kJ mol}^{-1}$ and so $K^\ominus = 1.53 \times 10^4$ at 1 000 K.

Hence, to conclude, in the temperature range considered both reactions have a similar type of behaviour but reaction 6 is always more favourable than reaction 5.

SAQ 6 (Objective 6)

The three main requirements are: (i) *selectivity* – the possibility of the occurrence of thermodynamically more favourable reactions (see Answer to Exercise 1 in Block 1) must be minimized by the use of a catalyst that specifically favours methanol production; (ii) *activity* – for reaction at any temperature, activity is essential, but for this *exothermic* process, a catalyst active at low temperatures would be a considerable advantage (Exercise 1 in Block 1 again); and (iii) *stability* – under high temperature and pressure conditions, the catalyst must maintain its activity by not sintering or becoming vitrified. It should also have an economically useful lifetime before becoming fouled or poisoned.

SAQ 7 (Objective 8)

(a) No. Physical adsorption, like the condensation of a vapour on to the surface of its own liquid, does not require an activation energy. This is clear from Figure 8, which shows that there is no 'energy hump' for a molecule to overcome as it approaches the surface of the substrate. (For porous catalysts the uptake of gases is often found to be diffusion-limited, so that the *overall* process of adsorption has an activation energy: but this is associated with the mass transport process and *not* the process of physical adsorption itself.)

(b) Yes. The process of desorption of a physically adsorbed species requires an activation energy. In Figure 8 it is clear that this activation energy is equal *in magnitude* to the enthalpy change for physical adsorption, ΔH_{pa}.

SAQ 8 (Objective 9)

The only way methane (and indeed other alkanes) can chemically adsorb is by a dissociative process, for example:

$$CH_4(g) + 2* \longrightarrow \begin{array}{c} CH_3 \\ | \\ * \end{array} + \begin{array}{c} H \\ | \\ * \end{array}$$

SAQ 9 (Objective 10)

(a) Call the activation energy for chemical adsorption of hydrogen molecules E_{ad}, and that for their desorption E_{des}. From Figure 10 (and the discussion of Figure 9), it follows that to desorb as hydrogen molecules the hydrogen must acquire energy equivalent to the magnitude of ΔH_{ca}, written as $|\Delta H_{ca}|$, plus the activation energy for adsorption, E_{ad}. Hence, $E_{des} = |\Delta H_{ca}| + E_{ad}$.

Given that $\Delta H_{ca} = -96\,\text{kJ mol}^{-1}$, then

$$E_{des} = E_{ad} + 96\,\text{kJ mol}^{-1}$$

As stated in the text, the value of E_{ad} is small – certainly less than $10\,\text{kJ mol}^{-1}$ – hence we would expect E_{des} to lie in the range 96 to $106\,\text{kJ mol}^{-1}$.

(b) The reaction of interest is:

$$\begin{array}{c} \text{H} \\ | \\ *\text{Ni} \end{array} \longrightarrow \text{H(g)} + *_{\text{Ni}}$$

If we use the notation introduced in Figure 9, then

$$D(\text{H}-*_{\text{Ni}}) = \tfrac{1}{2}(Z + Y)$$

where the label Z corresponds to $D(\text{H}-\text{H})$ and the label Y corresponds to the magnitude of ΔH_{ca}, $|\Delta H_{ca}|$. Thus

$$D(\text{H}-*_{\text{Ni}}) = \tfrac{1}{2}\{D(\text{H}-\text{H}) + |\Delta H_{ca}|\}$$
$$= \tfrac{1}{2}(432\,\text{kJ mol}^{-1} + 96\,\text{kJ mol}^{-1}) = 264\,\text{kJ mol}^{-1}$$

SAQ 10 (Objective 10)

The Lennard-Jones plot suggests that doping the iron surface with potassium has a marked effect on the physically adsorbed precursor state of *molecular* nitrogen. The enthalpy change for physical adsorption increases to $-48\,\text{kJ mol}^{-1}$, and there is a consequent reduction in the activation energy for dissociative chemisorption – in fact it becomes effectively zero. Note that this means the *rate* of dissociative nitrogen chemisorption is increased (at a fixed temperature) because of two factors: (a) the reduction in activation energy; and (b) an increase in the surface concentration of physically adsorbed molecular nitrogen as a consequence of its 'stronger' adsorption. Potassium clearly acts as a chemical promoter. In the industrial synthesis of ammonia, potassium oxide, K_2O, is used rather than potassium (Table 1), but it seems that the origin of the promotion effect is similar.

SAQ 11 (Objective 11)

Consider the following reaction:

$$\text{A(g)} + * \longrightarrow \begin{array}{c} \text{A} \\ | \\ * \end{array}$$

where A is a gaseous adsorbate that can be adsorbed associatively or dissociatively. If the reaction is to be thermodynamically favourable then $\Delta G_{ca} < 0$. However, since $\Delta G_{ca} = \Delta H_{ca} - T\Delta S_{ca}$, it follows that $\Delta H_{ca} - T\Delta S_{ca} < 0$, or $\Delta H_{ca} < T\Delta S_{ca}$. If ΔS_{ca} is negative, then for the inequality to hold it follows that ΔH_{ca} must be a larger, more negative quantity than the quantity $T\Delta S_{ca}$. Hence chemical adsorption is typically an exothermic process.

[Endothermic chemical adsorption can occur in certain circumstances. For example, J. H. de Boer demonstrated that if a molecule dissociates on adsorption to produce highly mobile species, then the value of ΔS_{ca} can be positive, in which event ΔH_{ca} can also be positive, as the inequality above indicates. Endothermic chemical adsorption has been observed when hydrogen is adsorbed on an iron surface contaminated with sulfide ions, and there is evidence that it occurs in some other systems.]

SAQ 12 (Objectives 10 and 12)

The differential heat of dissociative chemisorption of nitrogen on an iron surface, $q_d(\theta)$, should (and does!) depend critically on fractional surface coverage. We would expect it to *decrease* with *increasing* θ. With this in mind look at Figure 40, which shows a series of schematic Lennard-Jones potential energy plots illustrating how the variation of $q_d(\theta)$ will cause the activation energy for dissociative adsorption, $E_{ad}(\theta)$, to *increase* with increasing θ. It is interesting to recall from SAQ 10 that for nitrogen dissociation on iron the opposite effects, an increase in q_d and a decrease in E_{ad}, are brought about by the presence of potassium on the surface.

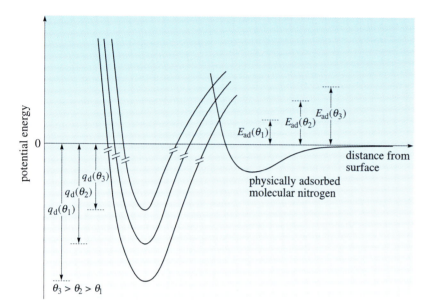

Figure 40 Schematic Lennard-Jones plots as a function of fractional surface coverage for the dissociative chemical adsorption of nitrogen on an iron surface. (For clarity, the 'mixing' of the physical adsorption and chemisorption curves in the region where they intersect has been omitted.)

SAQ 13 (Objective 13)

The peak γ in Figure 17 occurs at low temperature, approximately 150 K, and thus corresponds to a very weakly bound state with a small activation energy for desorption. The hydrogen is most likely physically adsorbed (as a molecular species) in this state. The peaks β_1 and β_2 correspond to sites that have much stronger binding energies. The chemical adsorption of hydrogen is normally dissociative and so it would seem that this particular crystal face of tungsten provides two energetically different sites on which this type of adsorption can occur.

SAQ 14 (Objectives 8 and 9)

Table 10 A comparison of physical and chemical adsorption.

Criterion	Chemical adsorption	Physical adsorption
(i) forces involved	chemical bond formation	weak intermolecular forces, relatively non-specific
(ii) enthalpy change for adsorption, ΔH	-40 to -400 kJ mol^{-1}	less than -40 kJ mol^{-1}
(iii) activation energy	usually small, but can be zero	zero
(iv) numbers of layers adsorbed	a monolayer, at most	multilayer adsorption possible

SAQ 15 (Objective 17)

The first step in the reaction (which has similarities to the adsorption of oxygen itself by a process such as $O_2(g) + 4e = 2O^{2-}(ad)$) will be favoured by a metal oxide that can provide electrons. The most efficient catalysts will therefore be p-type metal oxides. Notice that this conclusion assumes that it is the transfer of electrons to N_2O that is rate-limiting, which in fact seems to be the case.

SAQ 16 (Objectives 6 and 17)

A 'deep' oxidation catalyst is required. From a practical standpoint, it must be non-toxic(!), relatively inexpensive and water-insoluble. From a chemical viewpoint, it must be non-selective in action and able to promote the oxidation of various hydrocarbons and fats over a reasonable temperature range. These requirements point to the use of a p-type metal oxide. In fact, manganese oxide is often used, mixed with a frit (a mixture of silica, alumina and other components to give a glassy composition).

SAQ 17 (Objective 18)

If relative catalytic activity parallels the acid strength of active sites in the zeolites, then the most likely explanation for this behaviour is the operation of two opposing effects: increasing the Si : Al ratio *increases* the acid strength per site, but *decreases* the number of sites (which depends on the number of Al atoms in the framework). The first effect should dominate at low Si : Al ratios, when the density of sites is highest.

SAQ 18 (Objectives 6 and 18)

According to Table 4, zeolite CaA shows *reactant* selectivity, in that it cracks n-hexane, but not 3-methylpentane. Reference to Figure 30 suggests that this can be correlated with the pore structure: the windows in CaA admit straight-chain hydrocarbons, but exclude even the smallest branched alkane (*iso*-butane, for example). This also explains the observed *product* selectivity of this zeolite: according to the right-hand side of Table 4, branched products are essentially absent following cracking over CaA, whereas they are the prevalent products over amorphous silica–alumina. By contrast, zeolite NaA is not particularly active as a cracking catalyst, presumably because its smaller ports (Figure 30) severely restrict the entry and diffusion of both straight- and branched-chain hydrocarbons. This further highlights the 'fine-tuning' obtained by slight modifications to a zeolite framework – by ion-exchange, in this case.

SAQ 19 (Objective 18)

Two factors may be involved in increasing the selectivity for *p*-xylene. First, larger crystals have correspondingly longer pathways along which the products must diffuse out: the effect of differentials in the rates of diffusion of different products will thus be enhanced. Second (and you may well have missed this), selectivity is the result of reaction taking place at sites *within* the pore structure: larger crystals have a higher ratio of internal to external sites, thereby improving selectivity (since external sites are 'unconstrained'). Incidentally, this second factor can also be manipulated by irreversibly poisoning the external sites – by adsorbing molecules larger than the effective apertures of the zeolite, for example.

SAQ 20 (Objective 19)

Substituting $\theta = V/V_m$ into equation 40 gives

$$\frac{bp_A}{1+bp_A} = \frac{V}{V_m}$$

Multiplying both sides of this equation by V_m we get

$$\frac{V_m bp_A}{1+bp_A} = V$$

Multiplying both sides by $(1 + bp_A)$ gives:

$$V_m bp_A = V(1 + bp_A)$$

Dividing each term in this equation by V gives:

$$\frac{V_m bp_A}{V} = 1 + bp_A$$

Equation 42 can now be obtained by further dividing each term in this equation by bV_m:

$$\frac{p_A}{V} = \frac{1}{bV_m} + \frac{p_A}{V_m}$$

This is the equation of a straight line when p_A/V is plotted against p_A: the slope is $1/V_m$ and the intercept is $1/bV_m$. Thus the 'goodness of fit' of an experimental isotherm to the Langmuir model can be tested by seeing how good a straight line it gives.

Data for plotting $p(trans\text{-but-2-ene})/V$ against $p(trans\text{-but-2-ene})$ are listed in Table 11; the plot is given in Figure 41. This is a reasonably straight line, and so it seems that the adsorption of $trans$-but-2-ene on Bi_2MoO_6 at 298 K can be adequately described by a single-site Langmuir adsorption isotherm.

The slope of the straight line in Figure 41 is 9.87 cm^{-3}, so

$$1/V_m = 9.87 \text{ cm}^{-3}$$

and

$$V_m = 0.101 \text{ cm}^3 \text{ (at stp)}$$

The intercept of the line is 350 Pa cm^{-3}, so

$$1/bV_m = 350 \text{ Pa cm}^{-3}$$

Thus,

$$1/b = (350 \text{ Pa cm}^{-3}) \times (0.101 \text{ cm}^3)$$
$$= 35.35 \text{ Pa}$$

and

$$b = \frac{1}{35.35 \text{ Pa}} = 0.028 \text{ Pa}^{-1}$$

Table 11 Data for the adsorption of *trans*-but-2-ene on 1 g of Bi_2MoO_6 at 298 K.

$\dfrac{V(\text{stp})}{\text{cm}^3}$	$\dfrac{p(trans\text{-but-2-ene})}{\text{Pa}}$	$\dfrac{p(trans\text{-but-2-ene})/V(\text{stp})}{\text{Pa cm}^{-3}}$
0.005	1.6	320.0
0.011	4.4	400.0
0.021	9.3	442.9
0.037	19.3	521.6
0.052	37.3	717.3
0.066	68.0	1 030.3
0.073	92.0	1 260.3
0.080	133.3	1 666.3

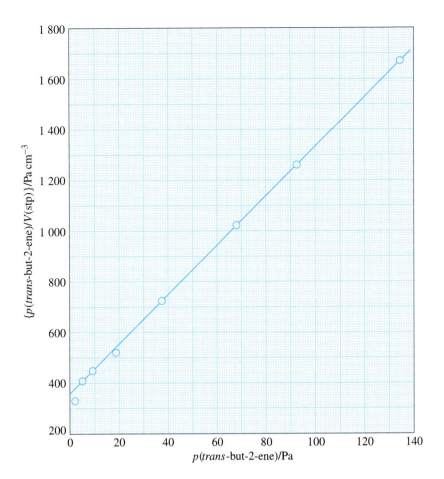

Figure 41 A test of whether the adsorption of *trans*-but-2-ene on Bi_2MoO_6 at 298 K can be described by a single-site Langmuir adsorption isotherm.

SAQ 21 (Objective 20)

According to equation 43,

$$b = b_0 \exp(-\Delta H_{ad}/RT)$$

Thus for the adsorption of *trans*-but-2-ene on Bi_2MoO_6 at 323 K:

$$b(323\ K) = b_0 \exp\{-\Delta H_{ad}/(R \times 323\ K)\}$$

and at 373 K

$$b(373\ K) = b_0 \exp\{-\Delta H_{ad}/(R \times 373\ K)\}$$

But $b(323\ K) = 7.95 \times 10^{-3}$ Pa^{-1} and $b(373\ K) = 1.01 \times 10^{-3}$ Pa^{-1}, hence

$$\frac{b(323\ K)}{b(373\ K)} = \frac{7.95 \times 10^{-3}\ Pa^{-1}}{1.01 \times 10^{-3}\ Pa^{-1}} = \exp\left\{-\frac{\Delta H_{ad}}{R}\left(\frac{1}{323\ K} - \frac{1}{373\ K}\right)\right\}$$

that is

$$7.871 = \exp\{-\Delta H_{ad} \times (4.992 \times 10^{-5}\ J^{-1}\ mol)\}$$

so

$$\Delta H_{ad} = -41.3\ kJ\ mol^{-1}$$

As expected, the adsorption is exothermic. The magnitude of ΔH_{ad} is, however, relatively low for chemical adsorption: but this is probably a result of 'fitting' the experimental data to the Langmuir model. Other evidence certainly suggests that *trans*-but-2-ene is chemically adsorbed on Bi_2MoO_6.

SAQ 22 (Objective 19)

For single-site adsorption,

$$\frac{p_A}{V} = \frac{1}{bV_m} + \frac{p_A}{V_m} \tag{42}$$

whereas for dissociative adsorption,

$$\frac{p_A^{1/2}}{V} = \frac{1}{b^{1/2}V_m} + \frac{p_A^{1/2}}{V_m} \tag{49}$$

Data for making the appropriate plots are listed in Table 12. Figure 42a shows the test for single-site adsorption, and Figure 42b the test for dissociative chemisorption. It is clear that the dissociative chemisorption isotherm fits the adsorption data better, that is it gives the better straight line. The slope of the line in Figure 42b is 9.84 cm^{-3}, so $1/V_m = 9.84$ cm^{-3} and $V_m = 0.102$ cm^3 (at stp).

Table 12 Data for testing whether the adsorption of *cis*-but-2-ene (CB) on Bi_2MoO_6 at 298 K is better represented by a single-site or a dissociative chemisorption isotherm.

$\dfrac{V(stp)}{cm^3}$	$\dfrac{p(CB)}{Pa}$	$\dfrac{p(CB)/V(stp)}{Pa\ cm^{-3}}$	$\dfrac{\{p(CB)\}^{1/2}}{Pa^{1/2}}$	$\dfrac{\{p(CB)\}^{1/2}/V(stp)}{Pa^{1/2}\ cm^{-3}}$
0.007	1.9	271.4	1.38	196.9
0.010	4.2	420.0	2.05	205.0
0.020	20.1	1 005.0	4.48	224.2
0.029	54.0	1 862.1	7.35	253.4
0.036	102.1	2 836.1	10.10	280.7
0.042	170.4	4 057.1	13.05	310.8
0.047	261.3	5 559.6	16.16	343.9

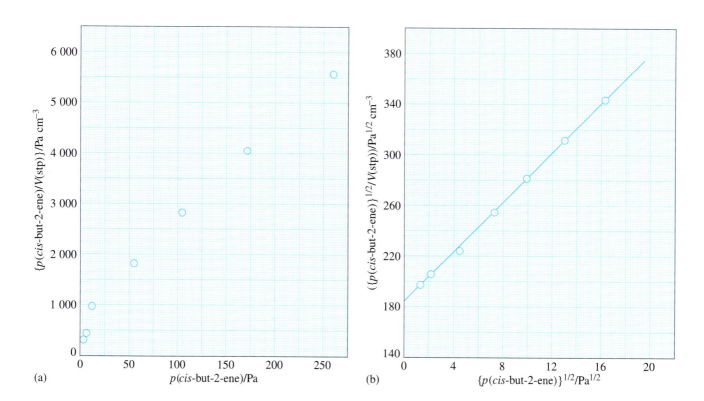

Figure 42 A test of whether the adsorption of *cis*-but-2-ene on Bi_2MoO_6 at 298 K can be described by (a) a single-site Langmuir adsorption isotherm or (b) a dissociative chemisorption isotherm.

The value of V_m is exactly the same as that for the adsorption of *trans*-but-2-ene on the same catalyst at the same temperature. It has been suggested that this indicates that there is a relation between the site on which the *trans* configuration adsorbs and that producing the adsorption of the *cis* isomer. The following mechanism has been tentatively put forward:

'butene' + $M^{n+} \longrightarrow$ π-bonded complex

π-complex + $2O_s^{2-} \longrightarrow (O-H)_{ad}^- + (O-allyl)_{ad}^- + M^{(n-2)+}$

where M^{n+} is a metal ion on the catalyst surface and O_s^{2-} is a surface oxide ion. The *cis* isomer performs both steps in the mechanism rapidly, whereas the *trans* isomer, perhaps owing to steric hindrance, is not broken down in the second step.

SAQ 23 (Objective 21)

The BET equation in the form of equation 57 is required:

$$\frac{x}{V(1-x)} = \frac{1}{CV_m} + \frac{(C-1)x}{CV_m}$$

For this equation: slope + intercept = $1/V_m$ (equation 58). Thus if the intercept is taken to be zero it follows that the slope is $1/V_m$. Using the one point on the plot, the slope can also be written as $x/V(1-x)$ divided by x, that is $1/V(1-x)$. Hence, for this single-point method, we have the simple relationship:

$V_m = V(1-x)$

$\quad = 140.4 \text{ cm}^3(\text{stp}) \times (1 - 0.25) = 105.3 \text{ cm}^3(\text{stp})$

$\quad = 105.3 \times 10^{-6} \text{ m}^3(\text{stp})$

Assuming ideal gas behaviour, the number of moles adsorbed is:

$$n_m = \frac{(101\,325\,\text{Pa}) \times (105.3 \times 10^{-6} \text{ m}^3)}{(8.314 \text{ J K}^{-1}\,\text{mol}^{-1}) \times (273.15 \text{ K})} = 4.698 \times 10^{-3} \text{ mol}$$

Thus, according to equation 56,

$S = (4.698 \times 10^{-3} \text{ mol}) \times (6.022 \times 10^{23} \text{ mol}^{-1}) \times (0.162 \times 10^{-18} \text{ m}^2) = 458.3 \text{ m}^2$

Thus the specific surface area is $458.3 \text{ m}^2/49.8 \text{ g}$, that is $9.2 \text{ m}^2\,\text{g}^{-1}$.

SAQ 24 (Objective 22)

In the first four rows of Table 7 the initial partial pressure of ethane is constant (0.03 atm). When the partial pressure of hydrogen is doubled, either from 0.1 to 0.2 or from 0.2 to 0.4 atm, the relative rate decreases by roughly a factor of four in each case. This strongly suggests that the relative rate is proportional to $p_{H_2}^{-2}$. Other combinations of values in the first four rows also support this conclusion.

In the last two rows of Table 7 the initial partial pressure of hydrogen is constant (0.2 atm). Increasing the partial pressure of ethane by a factor of ten increases the relative rate by the same factor. This strongly suggests that the relative rate is proportional to $p_{C_2H_6}$.

Thus the experimental rate equation is most likely to be of the form:

$$-\frac{dp_{C_2H_6}}{dt} = k_R \frac{p_{C_2H_6}}{(p_{H_2})^2}$$

that is $\alpha = 1$ and $\beta = -2$.

SAQ 25 (Objective 23)

For the reaction

$$C_2H_4(g) + H_2(g) \xrightarrow{\text{copper}} C_2H_6(g) \tag{64}$$

the Langmuir–Hinshelwood rate equation is given from equation 83 by:

$$r = -\frac{dp_{C_2H_4}}{dt} = \frac{k_\theta b_{C_2H_4} b_{H_2} p_{C_2H_4} p_{H_2}}{(1 + b_{C_2H_4} p_{C_2H_4} + b_{H_2} p_{H_2} + b_{C_2H_6} p_{C_2H_6})^2}$$

If C_2H_4 is strongly adsorbed compared with *both* H_2 and C_2H_6 (see for example Section 7.2), then $b_{C_2H_4}$ is much greater than both b_{H_2} and $b_{C_2H_6}$. The rate equation thus simplifies to

$$r = -\frac{dp_{C_2H_4}}{dt} = \frac{k_\theta b_{C_2H_4} b_{H_2} p_{C_2H_4} p_{H_2}}{(1 + b_{C_2H_4} p_{C_2H_4})^2}$$

If, in addition, the experimental conditions were such that $b_{C_2H_4} p_{C_2H_4}$ is much greater than 1, then

$$r = -\frac{dp_{C_2H_4}}{dt} = \frac{k_\theta b_{C_2H_4} b_{H_2} p_{C_2H_4} p_{H_2}}{(b_{C_2H_4} p_{C_2H_4})^2}$$

$$r = \frac{k_\theta b_{H_2} p_{H_2}}{b_{C_2H_4} p_{C_2H_4}}$$

This rate equation is of the same form as that determined experimentally at 273 K (with $k_R = k_\theta b_{H_2}/b_{C_2H_4}$).

If all three species, C_2H_4, H_2 and C_2H_6, are weakly adsorbed, then the general rate equation reduces to:

$$r = -dp_{C_2H_4}/dt = k_\theta b_{C_2H_4} b_{H_2} p_{C_2H_4} p_{H_2}$$

This is of the same form as the experimental rate equation at 473 K (equation 88), with $k_R = k_\theta b_{C_2H_4} b_{H_2}$.

Notice that it is reasonable that the adsorption coefficient $b_{C_2H_4}$ should *decrease* with increasing temperature (see, for example, Figure 35 and equation 43), and so cause the observed change in the form of the experimental rate equation.

SAQ 26 (Objectives 23 and 24)

(a) For a Rideal mechanism (see equation 89), in which an equilibrium concentration of molecular oxygen is established on the catalyst surface, the rate is

$$r = -dp_{O_2}/dt = -dp_T/dt = k_\theta p_T \theta_{O_2}$$

where p_T is the partial pressure of toluene. The equilibrium fractional surface coverage of molecular oxygen is given by (see equation 40):

$$\theta_{O_2} = \frac{b_{O_2} p_{O_2}}{1 + b_{O_2} p_{O_2}}$$

Hence

$$r = -dp_{O_2}/dt = -dp_T/dt = \frac{k_\theta b_{O_2} p_T p_{O_2}}{1 + b_{O_2} p_{O_2}}$$

(b) Let the fractional surface coverage of molecular oxygen at some time in the reaction be θ_{ox}, so that the fraction of sites vacant will be $1 - \theta_{ox}$. The rate of change of concentration of adsorbed molecular oxygen will then be:

$$\text{rate} = k_1 p_{O_2}(1 - \theta_{ox}) - k_2 p_T \theta_{ox}$$

In the 'steady state' the rate of change of concentration of the adsorbed oxygen will be zero, so that:

$$k_1 p_{O_2}(1 - \theta_{ox}) = k_2 p_T \theta_{ox}$$

or, on rearrangement,

$$\theta_{ox} = \frac{k_1 p_{O_2}}{k_1 p_{O_2} + k_2 p_T}$$

If the second step in the mechanism is taken to be rate-limiting, then

$$-dp_T/dt = k_2 p_T \theta_{ox}$$

Hence the overall rate of reaction can be written as:

$$r = -dp_{O_2}/dt = -dp_T/dt = \frac{k_1 k_2 p_T p_{O_2}}{k_1 p_{O_2} + k_2 p_T}$$

If an initial rate experiment is carried out with the initial partial pressure of oxygen ($p_{O_2,0}$) *fixed*, then the following expressions for the reciprocal of the initial rate of reaction can be derived for the two mechanisms.

Rideal mechanism

$$\frac{1}{r_0} = \frac{C_1}{p_{T,0}}$$

where $p_{T,0}$ represents the initial partial pressure of toluene and C_1 is a constant, given by

$$C_1 = (1 + b_{O_2} p_{O_2,0})/k_\theta b_{O_2} p_{O_2,0}$$

Two-step redox mechanism

$$\frac{1}{r_0} = \frac{C_2}{p_{T,0}} + C_3$$

where $C_2 = 1/k_2$ and $C_3 = 1/k_1 p_{O_2,0}$.

Clearly, a plot of the reciprocal of the initial rate of reaction against the reciprocal of the initial partial pressure of toluene will yield a straight line for both mechanisms. *However*, the plot for the Rideal mechanism will pass through the origin, whereas that for the two-step redox mechanism will have a distinct non-zero intercept. Hence, the detection of an intercept provides a means of distinguishing between the two mechanisms.

In fact, when this type of kinetic test is used, it is found that the Rideal mechanism can be discounted. However, the test is far from unique and cannot by itself be taken unambiguously to establish the two-step redox mechanism. Indeed, other quite plausible mechanisms have been proposed, which give rise to rate equations that fit the initial rate data reasonably well.

ACKNOWLEDGEMENTS

Grateful acknowledgement is made to the following sources for permission to reproduce material in this Block:

Figure 1 Philpott, J. E. (1971) 'Surface phenomena on rhodium–platinum gauzes', *Platinum Metals Review*, **15** (2) 1971, Courtesy of Johnson Matthey & Co; *Figure 3 Catalyst Handbook*, ICI Catalyst and Licensing Department and Wolfe Scientific, 1970; *Figures 4, 11, 19 and 21:* Bond, G. C. (1987) *Heterogeneous catalysis: principles and applications*, 2nd edn, Oxford Scientific Publications, by permission of Oxford University Press; *Figure 5* Courtesy of Prof. A. Datye; *Figure 6a* ICI Katalco, Billingham; *Figure 6b* Courtesy of Johnson Matthey plc, 1995; *Figure 12* Ertl, G. *et al.*, (1979) 'The role of potassium in the catalytic synthesis of ammonia', *Chemical Physics Letters*, **60**, (3), 1979; *Figure 14* Beeck, O. (1950) 'Hydrogenation catalysts', in *Discussions of the Faraday Society*, **8**, **1**950; *Figure 15* Ertl, G. *et al.*, 'Chemisorption of CO on the Pt(111) surface', *Surface Science*, **64**, 1977; *Figure 17* Madey, T. E. (1972) 'Absorption and displacement processes on W(111) involving CH_4, H_2 and O_2' *Surface Science*, **29**, 1972; *Figure 20* adapted from Somorjai, G. A. (1981) *Chemistry in Two Dimensions: Surfaces*, © 1981 by Cornell University, used by permission of the publisher, Cornell University Press; *Figure 26* Rabo, J. A. (1976) *Zeolite Chemistry and Catalysts*, © American Chemical Society; *Figure 29* Courtesy of Prof. J. M. Thomas, University of Cambridge; *Figures 30 and 31* Dwyer, J. (1984) 'Zeolite structure, composition and catalysis', *Chemistry and Industry*, April, 1984; *Figure 37* Everett, D. H. *et al.*, 'SCI/IUPAC/NPL project on surface area standards', *Journal of Chemical Technology and Biotechnology*, **24**, 1974, John Wiley & Sons Ltd., reprinted by permission of John Wiley & Sons Ltd.